Theoretical Methods in the Physical Sciences

to Bobbye
and to Evy and Kathy

William E. Baylis

Theoretical Methods in the Physical Sciences

*an introduction to problem solving
using Maple V*

Birkhäuser

Boston · Basel · Berlin

1994

William E. Baylis
Department of Physics
University of Windsor
Windsor, Ontario
Canada N9B 3P4

Library of Congress Cataloging-in-Publication Data

Baylis, William E. (William Eric), 1939-
    Theoretical methods in the physical sciences : an introduction to
problem solving using Maple V / William E. Baylis
        p.   cm.
    Includes index.
    ISBN 0-8176-3715-X (acid-free)   -- ISBN 3-7643-3715-X (acid-free)
    1. Physical sciences--Data processing.   2. Problem solving.
    3. Maple (Computer file)   I. Title.
Q183.9.B39   1994   94-3473
500.2'078--dc20   CIP

Printed on acid-free paper
© Birkhäuser Boston 1994

Birkhäuser

ISBN 0-8176-3715-X
ISBN 3-7643-3715-X
Typeset by the author.
Printed and bound by Quinn-Woodbine, Woodbine, NJ.
Printed in the U.S.A.

9 8 7 6 5 4 3 2 1

PREFACE

The advent of relatively inexpensive but powerful computers is affecting practically all aspects of our lives, but some of the greatest influence is being felt in the physical sciences. However, university curricula and teaching methods have responded somewhat cautiously, having only recently come to terms with the now omnipresent calculator. While many instructors at first feared that the widespread use of pocket calculators would lead to generations of students who could not multiply or perhaps even add, few now seriously lament the disappearance of slide rules, logarithm tables, and the often error-bound tedium that such tools of the trade demand. Time that used to be spent on the use of logarithm tables and manual square-root extraction can be profitably turned to earlier studies of calculus or computer programming. Now that the calculator has been accepted into the classroom, we face a computer-software revolution which promises to be considerably more profound.

Modern textbooks in the physical sciences routinely assume their readers have access not only to calculators, but often to home or even mainframe computers as well, and the problems teachers discuss and assign students can be more complex and often more realistic than in the days of only pad and pencil computations. As less effort is spent on numerical computation, more can be devoted to conceptual understanding and to applications of the increasingly sophisticated mathematical methods needed for a real appreciation of recent advances in the discipline.

While the calculator revolution was spreading through the schools, researchers in some branches of theoretical physics were developing programs to help them with the manipulation of algebraic and differential equations. Instead of computing numbers, such programs rearrange symbols and derive analytical solutions. The programs were initially bulky behemoths running only on large mainframes or supercomputers; they were also balky and awkward to use, and their capabilities were quite limited. Only dedicated scientists in specialized theoretical fields such as general relativity made much use of them. However, as personal computers have advanced to become more powerful than the supercomputers of two decades ago, the software has become faster and more encompassing, and an ever larger community of scientists has embraced the use of symbolic manipulation and computer algebra.

During the past couple of years, mathematics packages such as Maple, Mathematica, MathCad, Macsyma and Derive have become so common for PC-compatible, Macintosh, Atari, and other computers as well as for most Unix workstations that popular computer magazines, like *PC Magazine*, have featured and reviewed them. These programs generally combine abilities in symbolic manipulation with routines for numerical approximation and plotting. It seems likely that most students now entering university programs in the physical sciences will make some use of such mathematics packages before they receive postgraduate degrees.

The impact of these mathematics packages on university teaching may well prove far greater than that of pocket calculators. The packages will of course allow a wider variety of more complex, more realistic problems to be solved, but they will also inevitably change the way students approach problems. Instead of laboriously plotting data for a single experiment and fitting it to simple curves, much more data can be quickly visualized, fit and analyzed. Instead of grinding through the algebra to determine the coefficients of a power-series solution to a linear differential equation, the student can let the computer whip out the results and spend his or her time visualizing the results and relating them to physical phenomena. Instead of worrying about what substitution will simplify an integral to standard form so that its solution can be recognized, the student can concentrate on learning and applying new mathematical abstractions.

The introduction of powerful mathematics packages in the undergraduate curriculum will undoubtedly cause disruption and pedagogical debate. Just as when pocket calculators became common, there will again be fears about the loss of facility in mathematical manipulation. However, as the new tools become more popular, the need for such facility will decline, and the talents required of students making a career in mathematics may also change. The potential gains, including a greater appreciation of mathematical abstraction and hands-on experience with a broader selection of mathematical tools, will, once realized, far outweigh the losses.

The term *computer math* conjures up visions of number crunching on super-computers and mind-boggling numerical detail. Such aspects play an important role, but they are not the essence of physical science. Physical scientists make extensive use of simple models and pictures to convey concepts that are often more firmly based in geometry than in analysis. A course introducing theoretical methods in the physical sciences will naturally emphasize numerical and analytical techniques, but should not exclude the development of pictorial and geometrical concepts. Any math package chosen for the course should have superior graphical capabilities, and these should be used fully to present data in visual form.

This text grew out of a one-semester course which introduced a symbolic-mathematics package into the curriculum of the physical sciences at the University of Windsor. The objective of the course is not restricted to training students to use a symbolic math program; while that should be a practical by-product, the ability to tap artificial intelligence cannot replace the development of human wisdom. The most important skill science students can acquire is to learn science on their own through reading, deep thought and analysis, and discovery. No one would claim that this skill is easy to acquire: it requires concentrated effort on the part of the student, who can use all the help available. The math package is just one useful tool in guiding students to the new approaches and new thought processes needed by the physical scientist to increase her or his understanding of nature.

The text is designed for a one-term course, to be taken in the first or second year of university study. Since some familiarity with basic kinematics and elementary calculus is assumed, the course may be most effective after the student has successfully completed at least one semester of introductory courses in university-level physics and calculus. The reason for the early introduction of the material is to provide tools that the student can use throughout his or her career. The timing is critical since it is during the first year that many students decide (or their professors decide for them) whether or not they were cut out to be physical scientists. Their previous practice of memorizing equations and plugging in appropriate numbers, which served many quite well in high school, is no longer sufficient for academic success at the university. Now it is more important to relate equations to phenomena and mathematical abstractions to physical experience; still, this presents a major barrier to large numbers of students. Some of the tools and practice in this text, from dimensional analysis and order-of-magnitude estimates to Maple algorithms and geometric algebra, should help.

The course for which this text is designed is not meant to replace, and in fact is quite distinct from, the traditional third- or fourth-year course in mathematical physics. For example, although the symbolic math package learned here can be quite useful in solving differential equations and in studying special functions, such topics receive only a bare introduction on these pages. Instead, the text concentrates on areas where first- and second-year students in the physical sciences are known to have difficulty, either because of inadequate preparation or because of an underdeveloped problem-solving capacity.

The text may supplement—but should certainly not replace—the text in a general physics, astronomy, chemistry, or geology course; it makes no pretense of encompassing a systematic treatment of the first-year curriculum, and it lacks a laboratory of real physical experimentation which is so important for developing an "intuitive feeling" for the subject. Virtual experiments on a computer can supplement and extend physical experience, but they can never replace it.

This text is also not designed for a programming course. As an integral part, it uses the Maple package (currently Maple V, Release 3), an acclaimed leader in symbolic capabilities, developed since 1980 by scientists and mathematicians at the University of Waterloo, Ontario, the ETH in Zurich, and elsewhere. Extensive use is made of the symbolic-manipulation, numerical-approximation, and graphical capabilities of the package, but only minimal attention is paid to writing new procedures. Most of the examples are taken from elementary physics with some applications to astronomy, chemistry, and geology. The emphasis on physics reflects its role as the most fundamental and probably the most difficult of the natural sciences.

Geometrical concepts are emphasized in chapters 4 and 9 on vectors. Chapter 4 presents traditional vector manipulations of addition, scalar multiplication, and dot and cross products. Although vector components, basis vectors and reference frames are discussed, the stress is more on the concept of vectors as single entities. The relative position and motion of an object with respect to an observer, or of two objects with respect to one another, can be treated directly with vectors without the explicit introduction of components. Chapter 9 extends the vector concept to what is known as the geometric algebra of vectors and their products. It thus opens the rich potential of Clifford-algebra techniques to the reader.

It is assumed that students have access to the full Maple V package, Release 2 or higher, either on an independent PC or through a network connection, but most of the work discussed in these notes can be adequately carried out with the student edition of Maple V (Release 2 or higher), as well as on Unix or Apple Macintosh platforms. No prior knowledge of Maple or symbolic processing is needed, and indeed, this text can serve as an introduction to many features of Maple needed by physical scientists.

However, the present text makes no pretext of being a complete guide to Maple, and most serious students and researchers will want further information. A recommended source of additional information to Maple V, Release 2, is A. Heck, *Introduction to Maple* (Springer-Verlag 1993). The *Maple V Language Reference Manual* by B. W. Char, K. O. Geddes, G. H. Gonnet, B. L. Leong, M. B. Monogan, and S. M. Watt (Springer-Verlag 1991) is also a useful reference. The *Maple V Library Reference Manual* by Char et al. (Springer-Verlag 1991) is an optional reference of Maple procedures, although most of the information it contains is also available on-line through the Maple Help facility.

A $3\frac{1}{2}$-inch diskette has been packaged with the text for use with computers running Windows 3.1 or later. In addition to the Maple worksheets for each chapter, discussed below, it holds files for units, conversion factors and physical constants (Chapter 1), for properties of the chemical isotopes (Chapter 1), and for making calculations with the Pauli algebra (Chapter 9). Since the files are in seven-bit ASCII, it should be possible to transfer them and use them with Maple on other computer platforms. It is strongly recommended that the instructor supplement Chapter 8 on complex numbers with a copy of the freeware fractal program FRACTINT (for DOS or MS-Windows) or XFRACT (for Unix/X-windows). The program rapidly calculates and displays stunningly beautiful fractal drawings, explains the history and underlying mathematical algorithms, and allows users wide flexibility in the design and display of the images. Condensed files containing versions of the program may be downloaded from the Graphics Developers+ Forum (GO GRAPHDEV) of CompuServe,[1] and recent versions are also available on many bulletin boards and a number of FTP sites on the Internet, for example on SIMTEL20 and its mirror sites. Diskettes can also be ordered at nominal cost from shareware/freeware library services.[2]

About half of the sections of the current text contain examples and exercises to be tried on a computer. The headings of these sections begin with an asterisk (*). Much of this material, together with further exercises, is also available in the Maple worksheets which are stored as files on the diskette packaged with this text. There is one worksheet for each chapter, and although the text can be read without them, the worksheets do ease the students' introduction to Maple. Readers whose principal motive is to gain expertise on Maple may want to spend most of their time with the worksheets and use the text primarily as a reference. Most students, however, will probably want to progress one chapter at a time, first reading the text and then working through the worksheet on a computer.

Although many students will have some familiarity with computers, none is assumed, and a brief overview of the operation of stand-alone or networked PC's and the use of floppy disks is given in chapter 1 of these notes. Readers who are new to Windows may find it beneficial to go through the Windows tutorial so that they know how to manipulate windows, move through menus and select options, and cut or copy text with some combination of the mouse (or trackball) and the keyboard. No programming knowledge is needed.

It should be understood that reading in the physical sciences is very different from reading a novel. To be effective, it must not be a passive

---

[1] CompuServe is a registered trademark of CompuServe Incorporated, Columbus, Ohio. The file FRAINT.EXE is a self-expanding file which contains FRACTINT for MS-DOS systems, whereas the files WINFRA.ZIP and XFRACT.ZIP contain zipped versions of FRACTINT for MS-Windows and X-windows, respectively. All three files are located in the GRAPHDEV Library named "Fractal Sources." The current version number 18.2.

[2] For example, they are available from the Public (Software) Library, P.O. Box 35705, Houston, TX 77235-5705, USA. Phone (713) 524-6394 or (in the USA) 800-242-4775. Ask for item #9113.

process; it demands an active effort on the part of the reader. Here, the reader is encouraged throughout to think about what is being said and calculated, to relate results to physical experience, to make order-of-magnitude estimates of both answers and errors, and to carry out dimensional and unit checks of calculations. While standard problem-solving techniques such as "divide and conquer" are introduced, the theme is emphasized that there is usually no one correct method of solution, but rather many, and the more correct methods a student understands, the more insight will be gained into the problem, and the better she or he will be able to solve new ones.

Topics covered include the numerical approximation of real functions, data analysis and basic statistics, curve fitting, analytical and numerical integration and solutions to differential equations, complex numbers with elementary applications to fractals and chaos, and the (multi)vector algebra of physical space. Some topics recur in different guises. For example, the pendulum appears in Chapter 3 to relate uniform circular motion to simple harmonic motion, in Chapter 7 as a damped oscillator, and in Chapter 8 in complex solutions as an introduction to nonlinear motion. The coverage given each topic is necessarily restricted, and many topics such as Fourier analysis and Green-function techniques, usually treated in an upper level course in mathematical physics, are not treated at all. Maple, however, can contribute to their elucidation, and the instructor is encouraged to expand on areas of interest. Exercises, meant to be tried by the student as the material is worked through, are sprinkled throughout the text, and additional problems are collected at the end of each chapter. Some of the problems provide introductions to new areas; for example, a problem in Chapter 8 introduces coupled oscillators and eigensolutions. New terms and concepts are highlighted in italics and listed at the end of each chapter. The Maple worksheets mentioned above can help the student with some of the more difficult aspects of each chapter. New worksheets can be easily added.

Most instructors will probably find that the text contains somewhat more material than can be well covered in a single one-semester course. One possibility, certainly, is to work straight through the material and stop when the time is exhausted. Other options which still preserve the required development of ideas are to omit part or all of Chapters

5 and 6, Sections 4 through 7 of Chapter 7, and/or Sections 5, 6, and 9 of Chapter 8. More information about the interdependencies of the sections is given in the introductory sections of the individual chapters.

I am indebted to Edwin Beschler, Ann Kostant, and their associates at Birkhäuser Boston for their encouragement, helpful suggestions, and patience. I also gratefully extend thanks to all who read through early drafts of this text and offered suggestions for improvements, and I add a special note of appreciation to the wonderful trio of J's for their valuable work and good humor: J. Bonenfant, J. Derbyshire, and J. Huschilt.

Contents

Chapter 1

Introduction

The ability of scientists to judge sizes and recognize relationships is crucial to their discipline. It's an ability which would also serve the general public well in directing its energies most effectively toward such laudable goals as a cleaner and healthier environment. To take a contemporary example, it seems obvious that to gauge the severity of an accidental spill of radioactive material, it would be useful to compare the resultant radiation levels with the unavoidable levels arising from cosmic radiation. Yet how often have we seen the public attention in such cases focused on the total volume of the spill without a mention of the specific radioactivity, as though 100 liters of slightly radioactive water is much worse than 1 liter at a thousand times the concentration of radioactive isotopes? If radioactive levels are specified in news reports, how many persons can judge the significance without a reference to background levels? Understanding the significance of quantitative data is just one of the skills we associate with problem-solving in the physical sciences.

This course is an introduction to modern theoretical methods in the physical sciences. Its primary intent at a basic level is to teach problem-solving techniques, especially ones employing the Maple V software package in graphics and symbolic-mathematics. However, a higher-level goal is to train students in effective reading and learning strategies, in productive thought processes, and in sound approaches to scientific inquiry and research. It is emphasized at the start that there is no *single* method or technique for solving most problems. Like mountain summits to be scaled, most problems can be approached from different sides. A given approach can provide an easy but perhaps long journey to the peak, or it may be difficult and fraught with treacherous slopes. Which approach is best may be a matter of subjective judgment, since one may be the fastest, another may be the easiest, and yet another may reveal the beauty and symmetries of the problem most clearly. Superior students will often not be satisfied by simply scaling the summit from a single path; they will want to explore different approaches in order to learn as much about the terrain as possible. In the process of exploring the slopes, they will be preparing themselves for scaling new heights in the future.

Although most problems can be solved with a variety of approaches, some general principles can guide the explorer no matter which path s/he chooses. These general principles are the subject of Section 1.4 of this first chapter.

The chapter provides important background for the rest of the book. It begins by distinguishing physical sizes and numerical values. A simple but reliable method is reviewed for converting the units in which physical sizes are expressed, and the value of elementary dimensional analysis in checking equations is also discussed. Next, the chapter briefly discusses other principles, such as the divide-and-conquer and reduce-to-familiar-form techniques, and then it concentrates on order-of-magnitude estimates.

The second part reviews some of the basic information about the graphics computers to be used with this text and gives instructions for loading files from the diskette accompanying the text onto the computer hard disk. Those sections which are best covered in front of a computer are marked with an asterisk (*). The material of these sections which trains the reader in the use of Maple is also covered and extended in the Maple worksheets located on the diskette.

Students of this material should recognize that whereas gaining new understanding of the physical universe can be both great fun and greatly satisfying, it is rarely easy and it is never a passive process. No student should plan to read this text as if it were a novel. The acquisition of new wisdom requires the active participation of the student in formulating questions, anticipating answers, questioning, analyzing, and reviewing results, and discovering relations among bits of acquired knowledge. It takes time and concentrated effort, but the rewards make it all worthwhile.

1.1 Physical Sizes

How heavy is Fido? If he "weighs" 30 kg, we could also say he "weighs" 30,000 g, 3×10^{10} μg, 0.030 tonnes,[1] or about 66 lbs. Although the numerical values differ, all of these responses refer to the same physical size. The term "weighs" has been put in quotation marks because the physical characteristic given is actually a *mass M*, not a weight. Fido's true weight is his mass times the gravitational acceleration g of a freely falling object close to him. If Fido is on the surface of the earth where the gravitational acceleration is about 9.8 m/s^2, his weight must be

$$\text{Mg} = (30 \text{ kg})(9.8 \text{ m/s}^2) = 294 \text{ kg} \cdot \text{m/s}^2 = 294 \text{ N}, \qquad (1.1)$$

where 1 N = 1 Newton = 1 kg $\cdot$ m/s^2 is the metric unit of force.

There are a couple of conclusions to be drawn.

(1) If the physical size has dimensions, such as mass or force, its value cannot be given by a pure number. Rather, it must be a number of some appropriate units.

(2) In order to compare two values of a physical measurement, it may be necessary to convert from one appropriate set of units to another. The conversion should at most change the numerical value; it should not alter the physical value.

Of course the answers to some questions in mathematics and science *are* pure numbers. For example, "What is the value of pi to 50 decimal places?" or "how many micro lasers fit on the head of a pin?" In such cases, the correct answer is a pure number without units and no conversion can be done. However, when the answer has physical dimensions, it is meaningless to give a pure number without units. A statement such as "Fido weighs 30" conveys no useful information.

---

[1] The *ton* can be one of several units of roughly the same size. It may be the short ton (2000 lb), the long ton (2240 lb), or the metric ton (1000 kg). The unit *tonne*, on the other hand, is reserved for the metric ton.

1.2   Converting Units

Conclusion (2) above gives an important clue for unit conversion: to convert units of a given physical quantity, we can multiply by any factor which is a physical unity. Appropriate factors can be found from relations between units, for example

$$\begin{aligned} 1\text{ kg} &= 1000\text{ g} = 10^3\text{ g} \\ 1\text{ g} &= 10^6\ \mu\text{g} \\ 1\text{ kg} &= 2.2\text{ lbs} \\ 1\text{ tonne} &= 10^3\text{ kg} \end{aligned} \tag{1.2}$$

and so on. Factors of physical unity can be expressed in many ways:

$$1 = \frac{10^3\text{ g}}{1\text{ kg}} = \frac{1\text{ kg}}{10^3\text{ g}} = \frac{1\text{ tonne}}{10^3\text{ kg}} = \cdots \tag{1.3}$$

We look for factors that cancel the initial units and leave the final ones. Thus, to convert 30 kg to metric tonnes, we can multiply by physical unity in the form

$$30\text{ kg} = 30\text{ kg }(\frac{1\text{ tonne}}{10^3\text{ kg}}) = 3.0 \times 10^{-2}\text{ tonne}. \tag{1.4}$$

As another example, let's convert the average distance of the earth from the sun, which astronomers know as an *Astronomical Unit* (AU), to units of light-seconds. Given that 1 AU $\approx$ 150 Gm where 1 Gm = 1 gigameter = 10^9 m and 1 light-second = 299792458 m $\approx 3 \times 10^8$m, we find

$$1\text{ AU} = 150\text{ Gm}\left(\frac{10^9\text{m}}{1\,\text{Gm}}\right)\left(\frac{1\,\text{light-second}}{3 \times 10^8\text{m}}\right) = 500\text{ light-second}. \tag{1.5}$$

This means, of course, that it takes light 500 s or 500 s (1 min/60 s) $\approx$ 8.3 min to reach the earth from the sun.

> **Exercise 1.1.** Use factors of physical unity to convert your height from inches to meters and from meters to light-seconds.

Often the conversion of units is simple enough so that the systematic approach sketched here is not necessary. However, the technique of multiplying by physical unity is fast and simple and can avoid dumb but common errors such as dividing when you should have multiplied.

1.3 Elements of Dimensional Analysis

It is good practice to check every calculation for possible errors. Everyone can and does make mistakes, but the good scientist catches most of them before the results go to press. Of course, when you go to check a calculation, it's best to vary the method, since if you simply repeat the calculation in the same way, you are liable to repeat any mistake made the first time. Thus, for example, to check your addition of a column of numbers, you can try adding them in the reverse order or grouping the numbers in other ways.

If the calculated quantity has physical units, there is another valuable tool for verifying results: unit and dimension checks. Never *assume* that the units come out correctly: put them in and *make sure* they do! A simple check can reveal hidden dimensionless factors such as min/s or incorrect dimensions resulting from a more serious error in calculation. The effort required to check the units is usually insignificant compared to that for the calculation itself. If, for example, your calculation of the range of a projectile yields units of m^2 or $kg \cdot s$, there must be an error in the calculation: the result cannot be right since the range must have units of distance.

Dimensional checks can also be useful when the result is an equation or a formula instead of a number: simply plug in the dimensions of all terms and make sure that both sides of the result agree. If the dimensional check indicates a problem but no mistake is readily apparent, check the starting assumptions and formulas. Although good physics texts are generally more careful, it is unfortunately not uncommon for technical handbooks, engineering texts and others to give dimensionally unbalanced formulas which work if and only if numerical values corresponding to specific units are substituted for the variables.

For example, the *reverberation time* T_{rev} of a room of volume V, total surface area S, and average absorption coefficient[2] α, is the time for the

---

[2]The absorption coefficient is represented by the Greek letter alpha. You may have noticed a heavy use of Greek letters in the physical sciences. There simply are not enough letters in the common Latin alphabet to label all the important quantities. For that matter, there are not enough in the Greek and Latin alphabets *together* to label everything uniquely, but it helps to be able to use Greek letters. If you are not yet familiar with alpha to omega and everything in between, you may

acoustic energy in the room to fall by a factor of one million once the
source of sound is turned off. The absorption coefficient of a surface is
the fraction of sound energy striking the surface that is absorbed by it.
For a simple room, T_{rev} is given approximately by the Sabine formula:[3]

$$T_{rev} = 0.16\frac{V}{\alpha S}. \tag{1.6}$$

Because α depends on the frequency of the sounds present, so does T_{rev}.
(The formula overestimates the reverberation time at high frequencies
because it ignores absorption by the air itself.)

A quick check reveals dimensional nonsense: the left-hand side has
the dimension of time whereas the right-hand side, since α is dimensionless, has dimensions of volume over area, and hence of distance.
The formula does give the correct numerical value of T_{rev} in seconds
if V is the number of cubic meters of volume and S is the number of
square meters of surface area of the room, but if other units are used,
the result is garbage! In fact, the original Sabine formula is written[4]

$$T_{rev} = 0.049\frac{V}{\alpha S} \tag{1.7}$$

which works if and only if V is the numerical value of the volume in
cubic feet and S is the number of square feet in the surface area.

One way to correct such dimensional disasters is to give dimensions
to the numerical coefficient. Thus in 1.6, we could replace 0.16 by 0.16
s/m. Since 1 m $\approx$ 3.3 ft,

$$0.16 \text{ s/m} \approx 0.16\left(\frac{\text{s}}{\text{m}}\right)\left(\frac{1 \text{ m}}{3.3 \text{ ft}}\right) \approx 0.049 \text{ s/ft}. \tag{1.8}$$

so that the two Sabine *formulas* 1.6 and 1.7 are united in a single *equation*. A better approach is to use *dimensional analysis* and ask what
relevant physical quantity has dimensions of time divided by length.
Velocity has dimensions of length per time, so maybe the coefficient

find it helpful to spend a few minutes with the listing in Appendix A.

[3]For example, D. E. Hall, *Musical Acoustics* (Wadsworth 1980), p. 354.

[4]See for example J. Backus, *The Acoustical Foundations of Music*, second edition
(Norton 1977), p. 168.

is proportional to one over a velocity. What velocity? Well, the only velocity that could enter is the velocity of sound, or, more precisely, the *speed* of sound, and it makes sense that the time for the sound energy to decay is inversely proportional to the speed of sound: if the sound energy were to travel, say, twice as fast, it would strike the sound-absorbing surfaces twice as often. Since the Sabine formulas were determined close to room temperature, where the speed of sound is $v = 344$ m/s, we can replace the coefficient by 0.16 s/m $\approx 55/v$. The resulting equation

$$T_{rev} = \frac{55}{v} \frac{V}{\alpha S} \qquad (1.9)$$

is not only dimensionally balanced, it also expresses the dependence of the reverberation time on the speed of sound.

Simple dimensional analysis can thus indicate functional dependencies. A trivial example is the volume of a sphere. Since volumes have dimensions of $(length)^3$ and the only lengths associated with a sphere are the radius, the diameter, and the circumference, dimensional analysis suggests that the radius, diameter, and circumference are proportional to each other and that the volume V is proportional to the cube of the radius: $V = kr^3$. The analysis cannot provide the size of the dimensionless factor $k = 4\pi/3$ multiplying r^3 but can determine the rate of change dV/dr of the volume with respect to the radius: $dV/dr = 3kr^2$. Using the original relation $V = kr^3$ to eliminate k, we obtain

$$\frac{dV}{dr} = 3\frac{V}{r} . \qquad (1.10)$$

More generally, the dependence $y = kx^q$ yields the relation $dy/dx = qy/x$.

> **Exercise 1.2.** The average velocity $< v >$ of a viscous fluid through a length L of a cylindrical pipe is proportional to the pressure drop ΔP from one end to the other divided by L. It also depends on the radius r of the pipe and the viscosity η of the fluid. Use the dimensions of viscosity, namely, kg $\cdot$ m$^{-1\cdot}$ s^{-1}, to determine the dependence of $< v >$ on r and η.

Answer: $< v >= k \ (r^2/\eta)(\Delta P/L)$ where k is a constant, which turns out to be $1/8$. When the average velocity is multiplied by the cross sectional area, the volume flow rate $Q =< v > \pi r^2$ is obtained in a relation known as *Poiseuille's law:*

$$Q = \frac{\Delta P \pi r^4}{8 \eta L} . \tag{1.11}$$

Thus, for example, the flow rate increases by a factor of 16 when the radius is doubled.

1.4 Principles of Problem Solving

Much of physical science, and indeed most of physics is concerned with solving problems. Solving problems is rather like working out puzzles: you are rewarded with the satisfaction of finding the solution. However, unlike most puzzle solutions, the solution to a physical problem is usually meaningful in its own right and may lead to the discovery of new insights along the way. Furthermore, the best training for problem-solving is problem-solving itself. As discussed above, most problems can be solved in different ways. The solution which is most natural for you may appear awkward or hard to grasp for someone else, and *vice versa*. Once you have found a solution to a given problem, you can finally fully appreciate what the problem was. It is then useful to talk to other students about their solutions because you will be better able to understand their approaches and to assimilate new problem-solving techniques.

There are several principles which are generally useful for solving problems no matter what your favorite approach is. Here are ten such principles:

(i) **Understand the problem and cull the wheat from the chaff.** Some problems almost solve themselves once you have a clear picture of what is wanted. A sketch may help. Try to reformulate the problem in terms of explicit goals, and then consider what information you need. Remember that realistic problems often contain surplus information which can (and should) be ignored when you seek the solution.

(ii) **Be patient or even stubborn.** Most problems worth solving require analysis and thought. The more challenging the problem, the greater the reward in finding its solution and the more you can gain from it.

(iii) **Relate the problem to one whose solution is known.** Like all new knowledge, solutions to new problems are most easily comprehended if based on previously acquired information. You should be flexible when seeking relationships to other problems; particularly in the physical sciences, problems in entirely different fields often have almost identical solutions.

(iv) **Divide and conquer.** Some problems seem hard because they require many steps, even though each step is fairly easy to carry out. Wade in and start picking off one step at a time. You have probably used this technique to solve for the motion of compound Atwood machines.

(v) **View the larger picture.** As you divide to conquer, keep the final goal in mind and don't get side-tracked in a dead end. Sometimes a more general problem is easier to solve than one with specific conditions and values. The general picture may reveal some relationships and symmetries more clearly (see the next principle).

(vi) **Exploit the natural symmetry** of the problem to reduce the number of steps that have to be calculated independently.

(vii) **Choose simple notation.** The notation should be as simple as possible while emphasizing the symmetries and not being ambiguous. Why should you waste time writing down the same complex expression several times when you can always identify it with a symbol? (The value of good notation may be more important than generally recognized.)

(viii) **Keep track of approximations** so that you can verify their accuracy and estimate any error they may introduce.

(ix) **Verify that the results are reasonable.** Most problems in the physical sciences correspond to a real or imagined experiment, and while the results may well harbor surprises (that's part of the fun!), they should also be physically possible and, unless an approximation breaks down, even reasonable for physically realizable conditions.

(x) **Enjoy yourself.** This is like a game, remember? You will want to work hard, of course, but if you can also stay relaxed, you will be

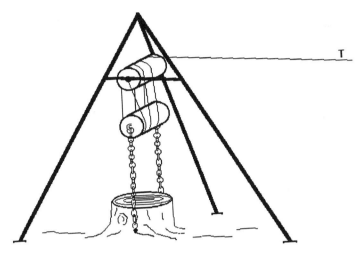

Figure 1.1. A block and tackle system for pulling tree roots out of the ground.

less likely to get stuck in a rut.[5] When some aspect of the calculation doesn't check out, try to avoid frustration by viewing the situation as an increased challenge.

As an example you may have seen before, consider the block and tackle rig pictured in Fig. 1.1 for pulling tree roots out of the ground.

A flexible, light-weight cable is wrapped over several groves of two pulleys, one of which is attached to the tree trunk while the other is suspended to a sturdy tripod. The problem is to find the maximum upward force which the rig can exert for a given tension in the cable. The problem looks messy: many loops of cable and a complicated object like the trunk and roots of a tree. However the solution is trivial.

We start by applying principle (i) and ignoring the complex shape of the tree roots; that's an irrelevant diversion. We could equally well be lifting a simple weight. We simply want to find the maximum force

---

[5]Like most good rules, this one has limitations: mathematical derivations require a clear head. The intoxicating effects of successful solutions should provide spirits enough, and other solutions of spirits probably won't help. One should heed the algebraists' admonition: Don't drink and derive!

F that can be lifted for a given tension T in the cable. The pulleys may be approximated as frictionless and the cable as weightless (principle viii). We can also ignore acceleration; it is enough if the tree roots are pulled up at a constant velocity.

It is important to understand what tension is. Here we can recall the simpler problem (principle iii) of a weight suspended from a single vertical cable. In that case the tension provides a force on the weight equal and opposite to the gravitational force. It also pulls the top of the cable down with a force of equal magnitude. In fact, since we are ignoring the gravitational force on the cable itself, any segment of the cable must be pulled with equal (magnitude of) force in opposite directions by the sections of cable on either side. If any section of the cable were replaced by a spring scale of the sort commonly used to weigh fish, the scale would read a weight equal to the magnitude of the tension in the cable.

Next we divide the problem into two parts (principle iv): how is the tension at one end of the cable related to that at the other end, and what is the total force on the lower pulley? By the argument in the previous paragraph, the tension everywhere along a massless cable is the same. The argument is not changed by the pulleys over which the cable passes, since they are frictionless and can add no force (only a change in direction). If one end of the cable is fixed to the top pulley housing and the cable loops around the bottom pulley n times, any closed, imaginary surface around the lower pulley would cut the cable $2n$ times. Therefore, the total force on the lower pulley is the weight F downward plus $2n$ times the tension T in the cable upward. Since the lower pulley is not accelerating, the sum of the forces must vanish, thus $F = 2nT$.

That's the answer. Let's check it (principle ix). It says, for example, that if $n = 3$, and if we can pull on the end of the cable with a force of 100 lbs, we should be able to apply an upward force of up to 600 lbs. That seems high, but if we consider how far we would have to pull the end of the cable in order to raise the weight on it a distance x, we see that we would have to pull a distance $6x$. Thus in this idealized case, the work we do in pulling the cable is equal to the work on the tree trunk.

Others principles of problem solving will be illustrated below.

1.5   Order-of-Magnitude Estimates

A good first step in any calculation is to make an order-of-magnitude estimate of the answer. Such calculations are also known as "back-of-the-envelope" calculations because they can usually be done rapidly in an informal setting and require little writing and no computers. In many cases, an order-of-magnitude calculation is sufficient; it may show, for example, whether or not a proposed experiment is feasible. In other cases, complications may pose enough uncertainty that an order-of-magnitude calculation is all that is practical.

An order of magnitude is usually taken to be a power of ten. Therefore, in order-of-magnitude calculations, factors of 2 or so don't matter; two such calculations that differ by a factor of two or three can be said to agree! Sometimes you will need to look up some basic data in order to complete a given calculation, and you should not hesitate to do so. However, with some ingenuity, you may be able to pool the relevant information in your head to come up with an order-of-magnitude estimate of what you need.

Suppose, for example, that we want to estimate the mass of the sun. Of course we could look it up, but that's no fun, and in any case, the table in which it's listed may have typographical errors in the stated powers of ten. What relevant facts do we know? We can figure out the acceleration of the earth toward the sun. Its magnitude is

$$a = \omega^2 r = v^2/r \,, \tag{1.12}$$

where $r = 1.5 \times 10^{11}$m is the radius of the earth's nearly circular orbit and $\omega = 2\pi/T$ is its angular velocity. Here, T is the period of the orbital motion, namely 1 year. The earth's orbital speed is $v = 2\pi r/T$. Newton's law of gravitation together with his second law of motion relates a to the mass M of the sun:

$$GMm/r^2 = ma = (2\pi/T)^2 mr \,. \tag{1.13}$$

Evidently the period is independent of the mass of the earth,[6]

---

[6]This relation embodies Kepler's third law: the square of the orbital period T is proportional to the cube of the mean orbital radius r.

$$GM = (2\pi/T)^2 r^3 \,, \tag{1.14}$$

but we don't know the gravitational constant G. Sure, we could look it up, but can't we estimate it well enough? The gravitational acceleration at the surface of the earth is similarly $g = GM/R^2 \approx 10 \text{ m/s}^2$, where $R = 6.4 \times 10^6$m is the radius of the earth. In terms of the average density ρ of the earth, $m = 4\pi\rho R^3/3$. The average density must be closer to that of nickel (about 9 Tonne/m^3) than to that of water (1 Tonne/m^3). An approximation of $\rho \approx 6$ Tonne/m^3 will be good enough for an order-of-magnitude calculation:

$$G \approx \frac{gR^2}{m} = \frac{3}{4\pi}\frac{g}{\rho R} \approx 6 \times 10^{-11}\frac{\text{m}^3}{\text{s}^2\text{kg}} \,. \tag{1.15}$$

Thus with $T = 365.25 \times 24 \times 60 \times 60 \text{ s} \approx \pi \times 10^7 \text{ s}$ we find

$$
\begin{aligned}
M &= \left(\frac{2\pi}{T}\right)^2 \frac{r^3}{G} \\
&\approx \left(\frac{2}{10^7\text{s}}\right)^2 \frac{(1.5 \times 10^{11}\text{m})^3}{6 \times 10^{-11}\text{m}^3\text{s}^{-2}\text{kg}^{-1}} \\
&\approx 2 \times 10^{30} \text{ kg}
\end{aligned}
\tag{1.16}
$$

The final result, by chance, is much more accurate than one should generally expect from order-of-magnitude calculations.

As another example, let's estimate the number of molecules ever breathed by Einstein that you inhale during a typical breath. This estimate requires a number of fairly crude approximations which we will want to check. The problem is complex, but we can solve it with the divide-and-conquer technique. First we estimate the number of molecules which ever made it to Einstein's lungs. You can estimate the volume of an average breath by breathing into a bag, and you can count the average number of breaths per minute. There is a considerable variation depending on the person and her/his activity, but I estimate a total exchange of roughly 10 liter/min or 0.16 liter/s as typical. Albert Einstein lived for about 76 years (from 1879 to 1955) and thus exchanged a total of roughly

$$(76 \text{ yr})(\pi \times 10^7 \text{ s/yr})(0.16 \text{ liter/s}) \approx 4 \times 10^8 \text{ liter} \qquad (1.17)$$

of air. Since one mole of an ideal gas contains Avogradro's number (about 6×10^{23}) of molecules and occupies 22.4 liter at standard temperature and pressure, each liter contains about 3×10^{22} molecules. Thus if we neglect the fact that some molecules were breathed more than once, Einstein must have breathed about 10^{31} molecules in his lifetime.

Of course, some of the molecules will have been in Einstein's lungs two, three or even more times. However, in any one breath, the fraction of "used" molecules is probably negligible and certainly less than 50% so that our estimate must give the right order of magnitude.

The next step is to estimate the total number of molecules in the atmosphere. Since atmospheric pressure of about $p = 100$ kPa $= 10^5$ N/m^2 measures the force per unit area and is a result of the weight Mg of the atmosphere spread over the surface area $4\pi R^2$, the total mass of the atmosphere must be

$$
\begin{aligned}
M &= 4\pi R^2 p/g \\
&\approx 4\pi (6.4 \times 10^6 \text{m})^2 (10^5 \text{ N/m}^2)/(10 \text{ m/s}^2) \\
&\approx 5 \times 10^{18} \text{kg} .
\end{aligned}
\qquad (1.18)
$$

Since the atmosphere is composed of roughly 80% nitrogen N_2 (with a molecular mass of 28 g/mole) and 20% oxygen O_2 (with a molecular mass of 32 g/mole), the average mass of the atmosphere is roughly 29 g/mole. The atmosphere therefore contains roughly

$$(5 \times 10^{18} \text{ kg}) \left(\frac{10^3 \text{g}}{\text{kg}} \right) \left(\frac{1 \text{ mole}}{29 \text{ g}} \right) \left(\frac{6 \times 10^{23} \text{molecules}}{\text{mole}} \right) \approx 10^{44} \text{molecules} .$$

Assuming that the 10^{31} molecules that Einstein breathed are evenly mixed among the 10^{44} molecules of atmosphere, about one in every 10^{13} air molecules once passed through Einstein's lungs, and a typical liter of air contains about $3 \times 10^{22}/10^{13} = 3$ billion such molecules.

Before leaving this inspirational calculation, we should consider some of the uncertainties. Some of the atmosphere present in Einstein's day may now be locked up in plant and animal tissue or in deposits of calcium carbonate and have been replaced by molecules from decaying organic matter and volcanic releases. Such effects could decrease the concentration of "Einstein's molecules" somewhat, but probably by less than a factor of 2. On the other hand, the circulation of the atmosphere is probably not complete, and as a result the effective pool of molecules for diluting those from Einstein may be smaller than estimated; the effect would now be to increase the concentration, but it is unlikely that the effect would change the calculated value by more than about a factor of two. We may thus have some confidence that the order of magnitude of our result is correct.

Perhaps you can see the power of order-of-magnitude calculations: with a little thought and understanding of basic laws you can piece together a tremendous amount of information. Details of the calculation depend on the knowledge at hand and may differ considerably from one person to the next. However, the results should agree within an order of magnitude, of course. Such estimates can be useful for checking the soundness of proposals and assertions, for determining the feasibility of an experiment or the veracity of a theory, and for finding a preliminary answer to a problem before undertaking a more thorough analysis. The ability of the physical scientist to make order-of-magnitude estimates is an underestimated talent which inspires awe in the uninitiated and can be a key element to the professional success of the practitioner. The best way to develop proficiency in the practice is to use it.

> **Exercise 1.3.** Estimate the total garbage (weight and volume) produced yearly by an urban population of one million people in North America.

1.6 *Getting at Maple

In this text, symbolic mathematics software by Maple[7] is used extensively to solve problems and plot results. In particular, Release 2, 3, or

---

[7]Maple is a trademark of Waterloo Maple Software.

higher of Maple V is assumed to be available to the reader. Maple V
has been produced for many computer platforms and this text should
be useful with most. However, most of the examples given assume
that Maple V Release 2 or later is running under the Windows[8] operating system on a PC-compatible computer. Most of you will have
experience with PC-compatible computers, but if you are new to the
computer age, don't worry. It will be fastest if you find someone to
help you get started, but you can also learn by trial and error. The
computer has the great virtue of patience, and you are unlikely to do
any serious damage as long as the important files are adequately backed
up by spare copies.

If you are unfamiliar with standard Windows operations, you may
find it useful to work through the Windows tutorial which comes with
most versions of the Windows operating system. The tutorial can be
selected from the Help menu of the Program Manager (see below for
more information about choosing menu selections). Review how windows are moved and sized and how the menus and items in them are
selected by the keyboard and/or the mouse or trackball. Whatever
version you use, it will also be useful to review the information on
files: how they are named, stored, deleted, copied, and retrieved. This
text will lead you through a bare minimum of the essential commands,
but it is no substitute for more thorough lessons on the operation of
computers and Windows.

The standard PC computers for use with Maple are 386 machines
or better. They typically have two floppy drives: drive a, for 1.2 MByte
5.25-inch disks and drive b, for 1.44 MByte 3.5-inch disks. Other
drives may also be available. A RAM (random access memory) of
two megabytes is barely adequate for running Windows; four, eight, or
more megabytes is better. A mouse (known as a *trackball* if it's on its
back) should be connected, and a printer may also be attached.

When your PC computer is first turned on, you may enter the Windows operating system directly; in that case you will probably see
a large window which contains other windows and group icons as in
Fig. 1.2 with Program Manager written at the top of the screen. The
group icons all have a similar appearance and look like miniature win-

---

[8]Windows is a trademark of the Microsoft Corporation.

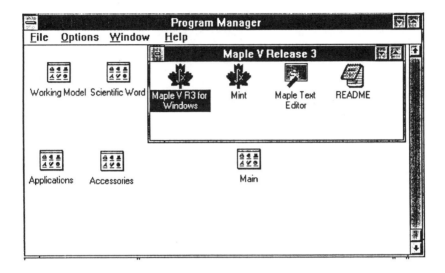

Figure 1.2. Windows screen with Maple V icon.

dows with six smaller icon images inside. You can open the window for a given group by moving the mouse so that its pointer is on the group icon and then "double clicking" the left mouse button, that is, pressing the button twice in rapid succession. In what follows, we will abbreviate this instruction to "double-clicking on the group icon."

If when you turn on your computer you are faced with a largely blank screen and a DOS prompt, which looks something like

`c:\>`

try typing `win` and pressing the <Enter> key (which on some keyboards is marked simply by a backward hooked arrow: ⟨←⟩) in order to bring up the Windows interface. On a networked system, you may need to *login* and type a *userid* and/or *password* before you can gain access to Windows or Maple. See your instructor or system administrator for details.

If you need to exit Windows and return to the DOS prompt, either press <Alt>-F (you hold down the <Alt> key while pressing the key with the letter F; this brings down the File menu) followed by X (this selects Exit on the File menu), or double-click on the control button at the top left-hand corner of the Window. If several windows are open, you may need to repeat this procedure for each. You will finally be asked to confirm that you really want to leave Windows, which you can do by pressing <Enter> or clicking the mouse on the $\boxed{\text{OK}}$ button.

Once you have an open Program Manager window, locate the Maple V icon, a maple leaf. It may be in its own window, as pictured in Fig. 1.2, or it may be in another window. You may have to open a few group-icon windows to find it. To start Maple, double-click on the maple leaf labeled "Maple V release n," where n is a number. If more than one release is available, choose the icon with the highest release number. To search through and manipulate files, double-click on the File Manager icon in the $\boxed{\text{Main}}$ window. Several windows can be open at once. You move from one window to another by clicking the left button of the mouse when it is pointing to part of the window to which you wish to move. If the window you want is hidden, you can reduce other open windows to icons by clicking on the downward-pointing triangle in the upper right-hand corner of the window. You can re-expand the window by double-clicking on its icon. It is often convenient to cycle through the active windows by pressing <Alt>-<Tab> (pressing <Tab> while holding down <Alt>). On some keyboards, the <Tab> key is simply marked by two arrows: $\boxed{\substack{\leftarrow \\ \rightarrow}}$.

Every window has a menu bar across the top, just below the banner with the window's name. To open a given menu, click the mouse on the menu's name or hold down <Alt> while pressing the key of the underlined letter in the name. Select an item in the menu by pressing the underlined letter or by double-clicking on the item. Close the menu by pressing <Esc> or by clicking the mouse somewhere away from the menu. The first menu on the menu bar is <u>F</u>ile and usually contains items to <u>O</u>pen, <u>S</u>ave, and <u>P</u>rint files, as well as an item to E<u>x</u>it from the application window.

> **Exercise 1.4.** Locate the switches and turn on your monitor and computer. Use the above instructions to open the File Manager in the $\boxed{\text{Main}}$ window. Expand the window

to full-screen size by clicking on the up triangle in the upper right-hand corner of the File Manager window. At the upper left, just below the menu bar, you should see a list of the available disk drives. The letters a and b usually label the floppy drives, and higher letters probably represent fixed or hard drives. Some of the drives, particularly on networked systems, may be *virtual* (which means they will disappear once the computer is turned off) or may exist on another machine (a file server), but these distinctions will be largely transparent (of no consequence) to us. The letter of the current drive will have a box around it. Different drives can be selected with the mouse. The directory tree of the current drive is shown to the left on the screen below. One of the directories in the tree will be highlighted, and the contents of that directory will be displayed to the right. A directory can contain other directories and files of various type, and these will be distinguished by the icons to the left of the file or directory name. By clicking the mouse on the arrow keys located on the *scroll bar* to the right, the displays can be scrolled up and down. Check out the available drives and their current directories. Open the menus in the menu bar to see what menu items are available. See if you can locate the Maple V directory, which may be labeled maplev*n*, and investigate its contents. To display subdirectories on the directory tree, double-click on the parent directory. Note that at the top of the directory display, Windows (like Unix) uses a double period (..) to represent the parent directory. When you are finished scanning the system directories, exit the File Manager and start Maple. Then enter "3 + 4*20 ;" on a line starting with the Maple command prompt (>) (don't forget the semicolon followed by the <Enter> key in order to get action out of Maple). Try some other arithmetic using the divide sign "/" and the exponentiation operator "^". If you make a mistake, correct it by using the backspace, cursor, and delete keys, or simply enter a semicolon, accept the error message, and try again. Then exit from Maple.

The Maple package may also be purchased for some home computers. Versions are available for 386 and 486 machines with co-processors (486-DX computers have co-processors built in) as well as for the Apple Macintosh and some Atari machines. A new student version of Maple V.2 for only about $100 was released in Summer 1993 for DOS- and Windows-based machines.

1.7   *Floppy Disks

Floppy disks are a convenient medium for saving information at the end of one session and introducing it at the beginning of another. Students must provide their own disks, which may be purchased in the campus bookstore or in most computer shops and stationery stores. Both 5.25-inch high-density double-sided (HDDS) diskettes, capable of storing at least 1.2 MBytes of data, and 3.5-inch high-density double-sided (HDDS) diskettes, capable of storing 1.44 MBytes, may be used. Lower-density disks, such as double-density double sided (DDDS) diskettes, can also be read and formatted, but information written to them is not always easy to read on low-density drives.

Data is stored as magnetic dipole moments in the thin magnetic layers on either side of the disk. The information can be lost if the disk is subjected to external magnetic fields or heat, or if the delicate surface of the disk is physically scratched or soiled. Data is written to the disks and read from them by small "heads" which ride within the diameter of a human hair of the rapidly spinning diskette surface. A scratch, dirt, or even the grease of a fingerprint cannot only destroy data on the disk, it can damage the drive which reads and writes to the disk. Proper care of your diskettes is therefore extremely important.

The 5.25-inch diskettes come in a soft plastic case with oval windows on either side through which the sensitive magnetic layer can be accessed. Be careful never to touch this surface. You should also avoid bending the diskette or writing on labels attached to its case with any implement other than a soft felt pen. Keep the diskettes at least a few inches away from any electronic device, including your monitor and the power supply of your computer, which may radiate stray magnetic fields. The 5.25-inch diskettes are inserted horizontally into the a:

drive with the label up and out. Once the diskette is in the drive, you must close the latch in order to allow the head to approach its surface. Release the latch to remove the diskette. When not in the drive, the diskettes should be kept in their specially lubricated paper sleeves.

The 3.5-inch diskettes are somewhat better protected by their hard plastic shells and the sliding metal door which covers the opening to their magnetic surfaces. They are inserted into the drive (usually drive b:) with the label either up or to the left and out. They should pop into place. To remove a diskette, press the button on the drive.

Unless the diskettes are marked "pre-formatted," they must be formatted once before use: insert the diskette in the drive, enter the Windows File Manager as described above, select the appropriate floppy drive (be sure you know the right drive!), and select Format disk in the Disk menu. Follow instructions on the screen. When you have finished formatting any diskettes you need, you can return to the File Manager by answering N <Enter> to the query of whether you want to format another disk.[9] Formatting prepares the surface by laying out the tracks (annular rings) and sectors where the data will be stored and by establishing a directory and file allocation table (FAT) to record where everything is stored. In the process, it erases any pre-existing data on the diskette. You only need to format a disk once; be sure you never format one that contains valuable data!

See the Windows Help menu toward the right of the menu bar for further details about window and file manipulation. Information on saving and reloading data from Maple sessions will be given in the next chapter.

1.8   *Worksheets in Maple V

When you start Maple V, you begin with a largely blank screen. The commands you type and Maple's responses create a *worksheet* of your

---

[9]Diskettes can also be formatted from the DOS command prompt with the instruction `format drv:` where `drv` is to be replaced by the appropriate drive letter. If the diskette is not high density, use `format a:/4` to format a 360 KB 5.25-inch diskette in drive a: and `format b:/t/80/n:9` to format a 720 KB 3.5-inch diskette in drive b:.

Maple session. Worksheets can be saved as files on disk and opened later for review or further work.

Maple worksheets have been prepared for every chapter of this text. They are stored as files in the `tmlib` directory of the diskette which was packaged with the book. Although the book can be used without them, the worksheets provide an easy tutorial for the reader in the aspects of the text which use Maple. Indeed, someone only interested in learning Maple can use the worksheets without the text for a speedy introduction to many of Maple's features.

To open a previously saved worksheet, start Maple and select Open in the File menu. The *file extension* of a worksheet is .ms (which stands for maple session), and any files with names that end in .ms in the current working directory are shown in the left half of the Open window. The current directory can be changed by double-clicking on a directory on the tree in the middle of the window. You should see the file-folder icon open when it becomes the current directory. The directories shown on the tree can be changed by selecting a different drive and by scrolling (use the mouse to "press" the arrows on the scroll bar to the right). When the desired worksheet is displayed, double-clicking on it will open it for your Maple session.[10]

The worksheets for this text have imaginative names like `ws1.ms` and such. There's one for this chapter. Why not give it a try!

For faster access, it is recommended that the `tmlib` directory be copied from the diskette onto the main Maple directory, which is named `maplev2` in Release 2 of Maple V and `maplev3` in Release 3.. The copy procedure, which also copies subdirectories and their contents, is easily performed from the File menu of the File Manager, located in the [Main] window.

1.9   *Unit Conversion on Maple

A file is available with conversion factors and physical constants for use on Maple. If you are using a copy of the diskette supplied with this text in drive `b:` of a PC system (it is strongly recommended that you

---

[10]Or highlight it by clicking on it once and then select [OK].

save the original diskette as a backup), the file can be read into your Maple session with the command

> read 'b:/tmlib/units';

Note that *back quotes*[11] are needed in order for Maple to treat the characters like / as part of the string of characters specifying the directory and file name, and that forward slashes (/) are used in Maple to separate directories (in place of the backward slashes [\] common in DOS commands).

It is more convenient if the entire tmlib directory is copied from the diskette to the main directory of Maple, which is probably called maplev$n$, where n is the release number, and if tmlib is made your current working directory. You can change the current working directory from within Maple by selecting Open in the File menu.[12] Use the scroll bar and disk selection if necessary to locate tmlib on the directory tree in the middle of the Open window. Double-click on it to open its file-folder icon. This indicates that it has become the current working directory (the file-folder icons for any subdirectories below tmlib should be closed). Return to the Maple window without opening a file by clicking on the cancel button or by pressing <Esc> twice. Now the command can be simply

> read units;

(Ask your instructor or system administrator for help, if you run into trouble.) Then you are ready for conversion!

The units have common symbols but usually end with (or contain) an underscore. Normally, units will be converted to metric quantities. For example, to convert 25 feet to meters, enter simply

> 25*ft_;

Maple should respond with the equivalent length in meters:

$$7.6200 \; m_-$$

If you wish to convert to non metric units, divide the quantity by the desired units. Thus to find the number of inches in a kilometer, enter

---

[11]On most keyboards, the back quote character is located on the key to the left of the number 1 key.

[12]You can also change the current working directory by selecting Save As ... in the File menu.

> km_/in_;

It is useful to think of the units as physical quantities themselves. Thus km_ = 1*km_ is the physical length of one kilometer, and the ratio above is one kilometer divided by one inch and gives how many times the length of one inch divides into a length of one kilometer. It gives the number of inches in a kilometer and NOT vice versa. Maple's answer is a pure dimensionless number:

$$39370.07874$$

Exercise 1.5. The horsepower (hp_) is a common unit of power for motor vehicles. Find its value in metric units.

Exercise 1.6. The calorie (cal_) is the amount of heat required to raise the temperature of one gram of water from 15C to 16C. Let Maple tell you how many Joules (J_) there are in one calorie.

Exercise 1.7. The metric unit for measuring fuel consumption of motor vehicles is liters (l_) per 100 km, which tells how many liters of fuel are needed to travel 100 km. Any vehicle which requires more than 10 liters/100 km is considered to be a "gas-guzzler" in some jurisdictions. The common English measure of fuel consumption is basically the reciprocal quantity, namely the distance traveled per unit volume of fuel. Determine the number of miles per U.S. gallon (gal_) which correspond to the "gas-guzzler" limit of 10 liters/100 km. Repeat for miles per imperial gallon (gal_imp). [Hint: Try (10*km_/l_)/ (mi_/gal_) for the first conversion. Why?]

The file units contains many of the common conversions used in the physical sciences. It also knows the metric prefixes Exa, Peta, Tera, Giga, Mega, kilo, milli, micro, nano, pico, femto, and atto. Any of these can be multiplied by any unit. Note that, in contrast to the units, the prefixes do not end in an underscore. To see what the prefix means, simply enter its name. For example, if you enter

> Tera;

Maple should tell you that Tera represents a factor of 1000000000000

(10^{12}); it is equivalent to the American "trillion." The prefix is often used to describe the beam energy of the latest particle accelerators: a TeV is the energy acquired by an elementary charge e when it is accelerated through a potential difference of 10^{12} Volts. To find the number of ergs in a TeV, type

> Tera*eV_/erg_;

Note, incidentally, that Maple will NOT understand the plural ergs_; the symbols for the units are usually in the singular form: erg_. See Appendix B for a listing of the file units. You can add more constants and conversion factors if you like.

A few of the common units include metric prefixes in the symbols. With the help of the file units, Maple will understand km_ to be equivalent to kilo*m_, kg_ to be the same as kilo*g_, and ml_ to be one milli*l_. Although metric prefixes like *centi* and *hecto* are not included, the units cc_, cm_ and hectare_ are recognized. The file also contains common physical constants such as c_ (the speed of light), hbar_ (Planck's constant divided by 2π), e_ (the elementary charge), m_e, m_p and m_n (the masses of the electron, proton and neutron), m_earth, m_sun and m_moon (the masses of the earth, sun and moon), R_earth, R_sun and R_moon (the radii of the earth, sun and moon), and r_earth and r_moon (the average orbital radii of the earth and moon). The value of r_earth is also known as an astronomical unit (AU_). To display it in terms of gigameters (1 Gm is a million km), enter

> AU_/(Giga*m_);

or, equivalently,

> r_earth/(Mega*km_);

To determine how long it takes light to travel from the sun to the earth, enter

> r_earth/c_;

and to express the results in minutes, enter

> "/min_;

where the double quote (") is Maple's shorthand for the last expression (see Chapter 2 for more details).

> **Exercise 1.8.** Accurate variations in the distance to the moon are measured by reflecting pulses of laser light from earth-based telescopes off reflectors placed on the moon by

astronauts between 1969 and 1972. How long should it take
a pulse to travel from the earth to the moon and back?
(You may ignore the radius of the earth in comparison to
the distance to the moon.)

Exercise 1.9. How far does light travel in one year? This
distance is known as a light-year. Express your result both
in kilometers and in astronomical units.

1.10   *Chemical Isotopes

A directory of reference information about isotopes of the chemical
elements is also contained on the diskette that accompanies this text.
In the following examples, we assume that the directory `tmlib` has
been made the current working directory of Maple V.[13] As discussed in
Section 1.9, the current working directory can be changed from within
Maple by selecting Open in the File menu and double-clicking on the
desired directory in the directory tree shown in the middle of the Open
window. Remember to back out of the File-Open procedure by selecting
cancel or by pressing <Esc> twice. To invoke the isotope information,
execute the Maple command

>   `read 'atom/atom.dat';`

This loads a table named `periodic` and reads the location of small
tables for each element. If you want to know the name of the element
with atomic number 11, type

>   `periodic[11];`

and Maple should respond with the name *sodium*. If you wanted to
know the name of the element whose chemical symbol is Hg, type

>   `periodic[Hg];`

and Maple should respond with *mercury*. If you want to know the
atomic number of mercury, you can type either

>   `mercury[Z];`

or equivalently

---

[13]If the directory 'atom' is not a subdirectory of the current working directory,
you will need to add its pathname to all of the `readlib` instructions in the file
`atom.dat`. This is easy to do with the replace option in an ASCII editor such as
`Edit`. Remember that Maple uses forward slashes to separate directories.

> `periodic[Hg][Z];`

Similarly, you can ask for the electron configuration (`elconfig`), average atomic mass (`A`) in u, based on a mass of 12 u for carbon-12, and the symbol (`symbol`) of any element.

In order to learn information about a specific isotope, you enter two parameters, the first being the isotope number. Thus to find out the nuclear spin of the mercury isotope with 199 nucleons, type

> `mercury[199,spin];`

In the same way, you can have Maple tell you the lifetime (`lifetime`) of radioactive isotopes (to be precise, this is the half-life), natural abundance (`abundance`), and atomic mass (`A`) of each naturally occurring isotope and a few artificially produced ones as well. Of course, you can use the results directly in calculations. For example, the molecular mass of a molecule of carbon dioxide CO_2 is

> `carbon[A]+2*oxygen[A];`

Since the files are in ASCII, it is easy to add isotope information, if you wish.

1.11   *Problems

1. The range of a projectile on a flat, horizontal terrain depends only on the angle at which it is fired, its initial speed v, and the acceleration of gravity g. Without looking up or deriving the equation for the range, use elementary dimensional analysis to decide how the range changes if the speed is doubled.

2. Estimate to an order of magnitude the number of water molecules on the earth and compare this to the number of air molecules found in Section I.5. State your assumptions and approximations clearly.

3. Estimate the average heat production in Watts of a person who must consume a net of about 2400 Calories per day in order to maintain his/her weight at a constant level. Recall that a dietician's Calorie is the same as 1 kcal = 4186 Joules and that 1 Watt = 1 Joule/s.

4. Estimate the total number of protons in the universe. You may assume that most of the matter in the universe is in stars in the form of hydrogen, that there are roughly 10^{12} galaxies in the universe, and that their average size is more or less that of the Milky Way, which has about 10^{11} stars whose average mass is about one solar mass.

5. How sensitive a scale is needed to detect the decrease in weight of a 1 kg mass when it is raised from sea level to an altitude of 4 km?

6. A *parsec* is the distance at which an object appears to shift its position by one second of arc with respect to the distant background when the position of the observer is moved a distance of one astronomical unit perpendicular to the line of sight. Use this definition and the conversions and constants in the file `units` to calculate the number of light-years in one parsec. [Hint: one parsec is 3600 AU/deg, where deg is a degree of arc, expressed in radians (given in the file `units` by `deg_`). Why? (How many seconds of arc are there in a degree?) Your final answer for the number of light-years should be a bit over $3\frac{1}{4}$. You may need to use the Maple command `evalf(...)` to evaluate numerical values of some expressions. See Section 2.5 of the next chapter for more details. Compare your result directly with the ratio of the values `pc_` and `ly_` in the file `units`.]

7. Estimate the rate of heat generated by fluorescent lights in a typical shopping mall with 200 stores. You may assume that all the electrical power consumed goes eventually into heat. Compare the result with the rate of heat generated by shoppers on a fairly busy day. (See also problem 3 above.)

1.12   Chapter Summary

1.12.1   Concepts

dimensional analysis

order-of-magnitude estimates

principles of problem solving

unit conversion (see also Appendix B)

metric prefixes

Greek alphabet (see Appendix A)

chemical isotopes

floppy disks and their use

hard drives

Windows

mouse and trackball

icons and group icons

directory trees

menu bar

scroll bar

File menu

Help menu

1.12.2   Maple commands

`read`

`evalf`

`+, -, *, /, ^`

`", ", "`

Chapter 2

Maple V for Physical Applications

This chapter is an introductory guide to some very useful and powerful software. Its purpose is to introduce students of the physical sciences to the symbolic manipulation program Maple, version V. It is assumed that Maple V has been properly installed and that the reader knows how to start or log onto the system (see Section 1.6 in the previous chapter). These notes are written mainly for Release 2, 3, or higher, operating on a system with a windows interface, especially on a DOS-based personal computer with a 386/387 or 486 processor or better and a color VGA or SVGA monitor, but most of the material is also applicable on an Apple Macintosh computer or on a system in a Unix environment with X-windows. Although many of the commands can also be used on the first release of Maple V, Release 2 or higher should be installed to use the worksheets. The Maple commands in this text and the accompanying worksheets have been tested on Maple V, Release 2 and Release 3, and it is anticipated that most will continue to work on later releases.

Maple can't replace thinking, but it can change what you have to think about. It can relieve much of the tedium of calculating sums or integrals, or of finding power-series solutions to equations. It is not only usually faster, but also less prone to error than humans. By freeing students and researchers from some of the nitty-gritty detail, Maple can encourage a better appreciation of the structure and physical sig-

31

nificance of the mathematics. Students gain a better understanding by seeing a greater variety of more realistic problems solved symbolically and by displaying many of the solutions in two- and three-dimensional plots.

The coverage in this and later chapters is not complete; there are many packages of tools that will not be discussed. The subjects emphasized here are calculus, linear algebra, geometry, two- and three-dimensional plotting, and statistics. However, conscientious users of this guide should become sufficiently skilled in the operation of Maple to extend their expertise to other subjects and to writing their own Maple procedures. See *Introduction to Maple* by A. Heck (Springer-Verlag 1993) and the *Maple V Language Reference Manual* (Springer-Verlag 1991) by B. W. Char et al. for those topics which are not included here and for more details on those that are. Be forewarned that Maple is a veritable cornucopia of information in mathematics, and a full understanding of all its features requires a formidable knowledge which most of us don't (yet!) possess. Do not be discouraged if you can't understand everything the first time through! Maple also has an extensive on-line help facility, invoked by selecting Help in the menu bar at the top of the screen by clicking on it with the mouse, by typing <Alt>-H, or by simply entering a question mark (?) at the beginning of a line. We'll discuss it in more detail below.

If this guide is to serve you well, you should be somewhat familiar with Windows and mouse operations on your computer. Review chapter 1 and Worksheet 1, and if you still feel insecure with Windows, work through the Windows tutorial, available through Help in the Program Manager . The asterisked sections of this chapter are meant to be studied with the assistance of a computer. All the examples should be tried: first make sure they work as advertised, then play with some variants; experiment with different numbers, different operations, and different procedures. Only with hands-on experience does anyone become proficient with computer applications. The worksheet for this chapter leads the user through most of the Maple material and provides additional practice. See instructions in Chapter 1 to start Maple and open the worksheet.

2.1 The Maple Method

As we saw in the last chapter, we communicate with the Maple "engine" in a worksheet, either one we open, like those on the diskette packaged with this text, or a new one. In the worksheet, we can toggle back and forth between *text* regions, containing comments which are ignored by the Maple engine, and *input* regions, comprising Maple commands. In this chapter, we are concerned with the input regions: how do we get the Maple engine to do what we want? We need to understand Maple's abilities and to learn how Maple interprets our commands.

Whereas traditional computer languages can store integers, floating-point numbers of fixed length, logical variables, and character strings for later manipulation, Maple can also store variable names and expressions containing names for manipulation. Among the expressions which Maple can store are equations and sets of equations with equal signs "=" relating two subexpressions. A simple example is Newton's second law

$$\texttt{force} = \texttt{mass} * \texttt{acceleration.} \qquad (2.1)$$

Note that the variable names can be quite long; in fact, they can be up to 499 characters long. This means, of course, that the expression $F = ma$ would not convey the force law because Maple would assume ma was a single variable. Instead, we must indicate the multiplication explicitly with an asterisk:*.

To store an equation, we must assign it to a name:

$$> \texttt{forcelaw} := \texttt{force} = \texttt{mass} * \texttt{acceleration;} \qquad (2.2)$$

An *assignment statement* always has this form:

$$> \quad < \texttt{name} > := < \texttt{expression} > ; \qquad (2.3)$$

In Maple, the symbol for the assignment relation, namely ":=", is NOT the same as an equal sign. An equal sign indicates a mathematical equality; it is a symmetric relation. Newton's second law, for example, could equally well be written

$$\texttt{mass} * \texttt{acceleration} = \texttt{force} \qquad (2.4)$$

and the effect will be exactly the same as above in (2.1). The assignment relation, on the other hand, always has the expression on the right and the name to which that expression is assigned on the left.[1]

Expressions can go on and on and occupy several lines. Maple uses a colon ":" or semicolon ";" to tell it when an assignment or other command is complete. Several commands can be strung together on a line; Maple won't act on them until it receives a colon or semicolon followed by the <Enter> key. Commands completed by a colon are acted upon, but no result is displayed on the screen. Commands completed by a semicolon are also acted upon, and in addition, the result is displayed on the monitor.

Of course, expressions can also be as simple as a constant. For example, an expression for vertical acceleration in the gravitational field at the surface of the earth might look like this:

$$> \quad \texttt{acceleration} \; := \; -9.8; \qquad\qquad (2.5)$$

It's not always convenient to feed dimensional units like m/s^2 to Maple. The user can assume that Maple does not often make mistakes, but s/he must be diligent to ensure that the data Maple works with give results in the desired units.

2.2   Tokens

When you press a letter, number, or symbol key on your computer keyboard, a code is sent to the computer. Each such key has its own distinct code, which can be represented by a decimal number between 0 and 255, or its octal (base-8) or hexadecimal (base-16) equivalent. The American Standard Code for Information Interchange (ASCII), in one of its variants, is the standard code widely used by computers around the world. There are also ASCII codes for the escape key <Esc>, the backspace key <←>, the tab key, and the space bar. The <Enter> or <Return> key usually sends a pair of ASCII codes, one representing

---

[1]This bears emphasis because in some languages, BASIC and Fortran, for example, all equal signs ("=") indicate assignments rather than mathematical equalities. The language of Maple is generally closer to standard mathematical usage than most programming languages.

a return to the beginning of the line and one representing a line feed. The control key <Ctrl>, the <Shift> key, and the <Alt> key send no ASCII codes themselves, but while they are pressed, they change the ASCII codes sent by the letter, number, and symbol keys. The other keys, namely the function keys <F1> through <F10> or higher; the cursor keys, including <End>, <Home>, <Pg Up>, and <Pg Dn>; and the insert key <Ins>, send several ASCII codes in rapid sequence.

Some ASCII codes can be sent by more than one key combination. For example, the backspace key often sends the same code as <Ctrl>-h, and the <Tab> key traditionally sends the same code as <Ctrl>-j. The codes can vary, however, depending on how the keyboard is "mapped" in the software.

How does Maple make sense of a sequence of keys which are pressed on the keyboard? It must take the string of ASCII codes received and *parse* it, that is, split it up into meaningful units. The meaningful units are called *tokens*. A token may be a natural integer, a punctuation mark, an operator, or a recognizable string of characters. These are summarized below:

- *Natural integer:* a sequence of one or more digits.

- *Punctuation Mark:* one of : ; ' ` | < > ( ) [ ] { }

- *Operator:* among others, one of the *binary operators* + - * / ^ ** = := < > <= >= <> -> . , .. and or union intersect mod, one of the *unary operators* + - ! \$ not . % , or a *nullary operator* " "" """"

- *String:* a letter or underscore followed by zero or more (up to 499 in Release 2, an arbitrary number in Release 3) letters, digits, and underscores and terminated by any other character, or any sequence of characters enclosed in back quotes (`).

We have already encountered the colon (:) and semicolon (;) punctuation marks which indicate the end of a command. The meanings of many of the other tokens is as you would guess, such as the binary arithmetic operators for addition, subtraction, multiplication, and division (+ - * /); exponentiation, for which you may use either a caret

(^) or a double asterisk (**); the relational operators for equality (=), inequality (<>), less than (<), less than or equal (<=); and so on. The use of many of these and other tokens will be illustrated later in these notes.

One token ends where another begins or where one of the *white-space characters* <space>, <Enter>, and <Tab> appear. Any number of white-space characters can appear together without affecting the result. Thus, the expression

$$A + B \tag{2.6}$$

has the same effect as

$$A \quad + \quad B \tag{2.7}$$

and

$$
\begin{array}{c}
A \\
+ \\
B
\end{array}
\tag{2.8}
$$

In the first case, the expression $A + B$ is parsed into tokens A, +, and B, because A is a string, whereas + is an operator and cannot be part of a string (unless the string is enclosed in back quotes). In the other cases, the separate tokens are recognized because of the intervening white-space characters.

In addition, there are four *escape characters*: ? , ! , # , and \, which instruct Maple to suspend its digestion of tokens. The question mark (?), when it appears at the beginning of a line, invokes the help facility. Letters between the ? character and <Enter> will narrow the search. An exclamation sign (!) at the beginning of a line is not accepted in version V Release 2 with Windows. (In earlier releases, it indicated that the rest of the line is a command for the computer, which for a DOS-based system means a DOS command; now the user can simply switch windows, say with <Alt>-<Tab>.) The pound sign (#), wherever it appears in a line, tells Maple to ignore everything that follows it in that line. It is useful for inserting comments for humans. See Appendix B for uses of the # sign in the file units. A backslash (\) by itself or a backslash followed immediately by an <Enter> is ignored by the parser. The backslash-<Enter> combination provides a convenient way to continue statements by splitting long tokens at the end of a line, whereas the backslash by itself can be used to divide long integers into shorter segments to make them more readable. Thus the natural integer 987654321234567890 can also be written 987\654\321\234\567\890. If spaces had been used in place of backslashes, the result would have been six tokens instead of one.

2.3 Maple Commands

The assignment statement (2.3) discussed above is one important command. Another is simply an expression by itself; the command tells Maple to evaluate the expression. An example, namely

$$> 3 + 4 * 20; \tag{2.9}$$

was given in Exercise 1.4 in the last chapter. In order to evaluate the expression, Maple does not simply take the operations in left-to-right order, but instead it computes the product first and then the sum: we say that the binary operator * for multiplication *has higher priority* than the operator + for addition. If we wanted Maple to perform the addition first, we would have to add parentheses:

$$> (3 + 4) * 20; \tag{2.10}$$

In order of decreasing priority, the binary mathematical operators are

$$(**, \hat{\ }), \ (*, \ /), \ (+, \ -), \ mod \ . \tag{2.11}$$

The highest priority binary operation is exponentiation, which is indicated either by a double asterisk (**) or a caret (^). The parentheses in the list (2.11) mean that the symbols enclosed have the same priority. Thus, multiplication (*) and division (/) have the same priority, as do addition (+) and subtraction (−). This is as it must be, since division by a number q is just multiplication by the *multiplicative inverse* q^{-1} of that number; similarly, subtraction by p means adding the *additive inverse* $-p$ of p. For the logical operators, and has higher priority than or, and the set operator intersect has higher priority than union. Unary operations are generally performed before binary ones. More examples are presented in the next section.

The help command, invoked by selecting Help in the menu bar or by typing a command line starting with a question mark (?), is useful to practically everyone who uses Maple.[2] To end the session and leave Maple, select File followed by Exit, or enter the command quit (alias done or stop). There are also commands to save data to files and read it back. By storing the current *environment* on a floppy diskette, a complex computation can be restarted later from the point at which it was interrupted. In addition, Maple accommodates conditional commands of the if...then...fi form and repetitive commands of the type for...do...od , as will be discussed later.

2.4   *Getting Started

To start Maple, select the Maple V icon on the window. If no window is currently present, either bring one up (by typing win on a DOS-based system) or type maple (and <Enter>) at the system prompt[3] on other computer platforms. This should load the Maple executive procedure from hard disk into the computer's RAM (random access memory) and present you with a nearly blank worksheet. You can start calculating something right away.

To use Maple as an expensive calculator, try

$$> 5 * 20 + 50/5; \tag{2.12}$$

Don't forget to type the semicolon followed by the <Enter> key; that is what tells Maple to act. If you make a typing error, just backspace over it and retype. You can also move forward and backward in the line without erasing by using the cursor keys $\rightarrow, \leftarrow$, *home* and *end*. In this case Maple should return the value 110 because multiplication and division have higher priority than addition and are thus performed before it. Parentheses can indicate a different order of calculation:

$$>   5 * (20 + 50)/5; \tag{2.13}$$

---

[2]In a DOS system, the help command is also available by pressing the F1 key (see Section 2.6).

[3]Try xmaple on an X-terminal system running under Unix.

should yield 70. Since Maple ignores spaces between terms and operators in an expression,

$$> \ 5 \ * \ (20 \ + \ 50) \ / \ 5; \qquad (2.14)$$

should give the same result, namely 70. You might also compare

$$> \ (2\hat{\ }3)\hat{\ }4; \qquad (2.15)$$

with

$$> \ 2\hat{\ }(3\hat{\ }4); \qquad (2.16)$$

to show that exponentiation is not associative. That's why the parentheses are necessary here. The expression

$$> \ 2\hat{\ }3\hat{\ }4; \qquad (2.17)$$

is undefined and causes Maple to indicate an error. Go ahead and try it. You can use ** in place of ^ to indicate exponentiation if you choose. Try a few more examples of your own invention. What happens if you calculate with real numbers instead of integers?

> **Exercise 2.1.** Addition and multiplication are commutative: they do not depend on the order of the factors. Thus for example, $2 + 10 = 10 + 2$ and $2 * 10 = 10 * 2$. What about exponentiation? Let Maple compute 2^4 and 4^2 and compare the results. Do 2 and 4 commute under exponentiation? Does that mean that all integers commute under exponentiation? Try 2^10 and 10^2 as well as some other combinations of your choosing.

2.5 *Example - Radioactive Decay

Let's determine what fraction of the uranium-235 present at the formation of the earth 4.5 billion[4] years ago is still around. You may recall that this fraction is given by

$$F(t) = \exp(-\Gamma t) \qquad (2.18)$$

---

[4]This is the American billion, i.e. a thousand million and hence 10^9.

where Γ is the decay rate. As we found in the worksheet for Chapter 1 from the reference file on chemical isotopes, the decay of U-235 has a half-life of about 0.704 billion years. That means, of course, that when $t = t_{1/2}$ 0.704 GY (GY means gigayears), $F(t)$ is one half:

$$F(t_{1/2}) = 0.5 \qquad (2.19)$$

From this we should be able to determine the decay rate Γ, but let's turn the job over to Maple. Define the decay function by typing

$$> \; \texttt{F} \; := \; \texttt{t} \; -> \; \texttt{exp}(-\texttt{Gamma} * \texttt{t}); \qquad (2.20)$$

to assign the name F to the function that follows the assignment symbol := . In mathematical jargon, we say that the function is defined by the mapping (from the real numbers to the real numbers) of t onto $\exp(-\Gamma * t)$. Be careful to get the capitals right: unlike DOS, Maple is sensitive to the case. Thus exp, Exp, EXP are all different. We want to specify the exponential function exp, that is, a power of the number $e = 2.71828....$ Similarly, Gamma $\neq$ gamma (a predefined constant equal to 0.5772...) $\neq$ GAMMA (a built-in function). If you remembered to type the semicolon at the end of the assignment statement, Maple will have echoed your assignment.

To get Maple to solve for Gamma when F(t_half)=1/2, type

$$> \; \texttt{solve}(\texttt{F}(\texttt{t\_half}) = 1/2, \texttt{Gamma}); \qquad (2.21)$$

(The second character in the variable name t_half is an underscore; recall that it is a valid string character.) Maple's solution is

$$\frac{\ln(2)}{\texttt{t\_half}} \qquad (2.22)$$

which we can assign to Gamma by the instruction

$$> \; \texttt{Gamma} \; := \; "; \qquad (2.23)$$

As mentioned above, for U-235, t_half = 0.704:

$$> \; \texttt{t\_half} \; := \; 0.704; \qquad (2.24)$$

Finally, we ask Maple to evaluate the fraction F at t = 4.5:

$$> \quad \texttt{evalf(F(4.5));}$$
$$.01190700634 \tag{2.25}$$

In other words, just a little more than 1% of the radioactive uranium-235 remains of the original amount present when the earth was formed four and a half billion years ago.

> **Exercise 2.2.** What was the fraction of U-235 remaining 100 million years after the formation of the earth? after 2 billion years? What fraction will exist when the earth is twice as old as it is now?

Obviously this rather simple calculation didn't require the power of Maple, and indeed, the example barely hints at Maple's considerable capabilities. Nevertheless, it's a good starting point. Let's briefly review the steps so that you understand what happened and how you might extend this sort of computation.

We made several *assignments*, assigning a functional expression to the name F and assigning constants to variables. As discussed in Section 2.1, each assignment command has the same form:

$$< \texttt{name} > := <\texttt{expression}>; \tag{2.26}$$

with the relational operator := relating a name on the left with an expression on the right. A constant like 0.704 is just a simple expression. As with most commands to Maple, the assignment statements end with a semicolon (;). As mentioned above, we could also have used a colon (:) instead of a semicolon. Maple still accepts the statement and performs the calculation, but its response changes. Do you remember what the response should be? Well, try it!

It's easy to forget the colon or semicolon, but Maple doesn't know you're finished until you send it. Don't worry if you hit the <Enter> key before the colon or semicolon, Maple just sits there waiting for you to finish. Since the computer has infinite patience (except in lightning storms when it may lose power and suffer memory loss), you must yield: just type

$$; < \text{Enter} > \tag{2.27}$$

and Maple will jump to action.

The first assignment we made is a function definition. Maple is a symbolic manipulator, and such assignments have the same form as when a variable is assigned a constant value. One use of symbolic manipulation was seen when you solved for Gamma: Maple gave an expression which showed that Gamma is inversely proportional to the half-life t_half. Most programs would have been able to give us Gamma only when t_half was known numerically, and they would have given a numerical result for Gamma. With such programs we could easily have missed the functional relationship between Gamma and t_half.

Maple likes to be exact; if the numbers you give it are exact integers or rational numbers, it resists making numerical approximations until we ask for it. Try[5]

$$\begin{array}{ll} > & \texttt{sqrt(2); Pi;} \\ > & \texttt{sqrt(2.0);} \\ > & \texttt{sin(1/5);} \\ > & \texttt{sin(0.2);} \end{array} \tag{2.28}$$

We can find a numerical approximation by specifying the `evalf()` function, which tells Maple to **eval**uate a **f**loating-point value. Try the statements

$$\begin{array}{ll} > & \texttt{evalf(Pi);} \\ > & \texttt{evalf(sqrt(2));} \\ > & \texttt{evalf(sin(1/5));} \end{array} \tag{2.29}$$

Of course we neglected to tell Maple that our value for the age of the earth was valid only to two digits. Most of the 10 digits it gave us for F in (2.25) were not significant. Before we evaluated F we could have assigned

$$> \quad \texttt{Digits := 3;} \tag{2.30}$$

Then the calculation of

$$> \quad \texttt{evalf(F(4.5));} \tag{2.31}$$

---

[5]A list of common functions in the form that Maple recognizes is presented in Appendix C.

would have been rounded off to three digits. (It's generally a good idea to keep one more digit than the least accurate factor in the calculation.) Try it. Digits is a Maple environment variable which determines the accuracy of Maple's floating-point evaluations. If it is not set manually, it assumes the default value `Digits := 10`. Also try

$$
\begin{aligned}
&> \quad \texttt{Digits := 5;} \\
&> \quad \texttt{evalf(sin(2));} \\
&> \quad \texttt{evalf(Pi,40);}
\end{aligned}
\qquad (2.32)
$$

The optional second argument in `evalf()` temporarily overrides the value of `Digits`. Incidentally, note that the arguments of all trigonometric functions are assumed to be in *radian* measure, where π radians $= 180^o$. To check, try

$$
> \quad \texttt{sin(Pi/2); sin(Pi);} \qquad (2.33)
$$

These do not require the `evalf()` command because their results are integers and can therefore be given exactly.

2.6   *Help

The Maple screen is not quite empty when a new worksheet is first opened. Of the various menus on the menu bar at the top of the screen, one of the most useful is the Help menu at the right. If you select it, either by clicking on it or by pressing <Alt>-H, you will see a short menu offering both a Browser and Interface help. The current status of the system is found elsewhere: the key F2 gives the amount of memory used and remaining and the CPU time used in Release 2; in Release 3, the information is given at the bottom of the Maple window.

The F1 key gives an additional way to access the browser help; <Alt>-F1 gives you interface help. Start by selecting Interface help. A new window should appear with instructions and a list of underlined topics. Note that when the mouse pointer is moved to the list of topics, it changes to a pointing finger. The finger can be used to select one of the topics for more information. You can scroll through the Help window by clicking on the scroll-bar arrows, dragging the slider on the scroll bar, or by pressing the *PgDn/PgUp* and cursor-movement ↑ / ↓ keys. Select *Working with Work Sheets* and note how to open Maple's *Quicktour Worksheet* for a quick overview of Maple's abilities. Give it a test drive. Next select Browser and scan through some of the available topics.

Another way to call for help is type a command in your worksheet by starting the line with a question mark (?). To jump to a particular topic, enter "?" followed by the topic of interest:

> ?<topic>

If you misspell the topic, Maple may suggest alternatives. The help command (?) followed by a single lower-case letter will produce a list of possible topics starting with that letter.

For example, after concluding a command with a semicolon and <Enter>, type ?diff to learn about the differentiation operator. Press <Enter> to display information, and then use the cursor keys to move down to the examples. Select the line with the third derivative of sin$x$ by holding down the <Shift> key while you move the cursor across it, or drag the mouse across it. With that line highlighted, press <Ctrl>-C to copy it. Click on the Worksheet window and locate the cursor at the beginning of a command line. Then press <Ctrl>-V to paste the example. To run the command, then press <Enter>. If the result is another executable command, enter the command

> ";

Exercise 2.3. There's another differentiation operator called simply D. See what you can learn about it.

A good way to learn more about Maple and some of its capabilities is to browse through the help menus and screens. In fact, for many of Maple's tools, unless you get a large reference book like the *Maple V Library Reference Manual,* it's the only way to learn about them: there's much more information on-line in the help screen than in the language reference manual. Again, don't worry if you don't understand many of the mathematical packages and functions; some of the stuff is pretty obscure! You can learn more about them later if you need to.

Many of the help screens contain examples that you might like to run. Another way to see some of Maple's examples is with the `example` command. You can read about it in the help menu and run some of example's examples.

Try the commands

$$\begin{array}{ll} > & \texttt{example(diff);} \\ > & \texttt{example(D);} \end{array} \qquad (2.34)$$

to see what you get. We will discuss the differentiation operators later in this chapter. In any time remaining, tour the other topics to see what's available. To quit Maple, press the <Alt>-F, X combination and then confirm your action in a *dialog box*.

2.7 Simple Functions

The exponential function $\exp(x)$ used in Section 2.5 is an example of a simple function. It takes any real number x, its argument, and "maps" it to another real number, namely to the function value $f(x) = \exp(x)$. We write $f : x \mapsto \exp(x)$. The *domain* of the function is the set of possible values of the argument. In this case, the argument x can be any real number; therefore the domain of $f(x) = \exp(x)$ is $\mathbb{R}$, the real numbers. The *range* of the function is the set of possible values of the function $f(x)$ itself, in this case $\mathbb{R}_+$, the positive real numbers. For each value of the argument, there is only one function value $f(x)$. The exponential function is "simple" in the sense that both the argument and the function values are single real numbers. Later we will meet more complicated functions where the argument, and perhaps the function itself, is a position in three-dimensional physical space. Such positions can be represented by a list of three real numbers.

The simplest functional dependence is no dependence at all. If we ask, for example, how the magnitude of the gravitational acceleration at the surface of the earth changes in time, the answer, as far as one can tell, is that it doesn't change. The corresponding functional relation can be written

$$g(t) = g(0) \approx 9.8 \text{ m/s}^2. \qquad (2.35)$$

This is a valid function, even if it is rather boring. Its general form is

$$f(x) = a \qquad (2.36)$$

where a is a constant. A slightly more interesting functional dependence is *linear dependence*. The restoring force of a spring on a mass which is displaced by a small distance x from its equilibrium position is linear:

$$F(x) = -kx \,, \qquad (2.37)$$

where k is the spring constant. The more general form of a linear expression is[6]

$$f(x) = a + bx, \qquad (2.38)$$

where both a and b are constants. A *quadratic functional relationship* would add a term proportional to x^2 : $f(x) = a + bx + cx^2$, while a *cubic functional relation* would add another term proportional to x^3 : $f(x) = a + bx + cx^2 + dx^3$.

The exponential function is also quite common. It arose in our example of radioactive decay in Section 2.5. Recall that the fraction F of the initial radioactive material which still remains after a time t is given by the function

$$F(t) = \exp(-\Gamma t) \,. \qquad (2.39)$$

If there are $N(t_o)$ atoms of radioactive substance at the initial time t_o, then at a later time $t + t_o$, the number remaining is

$$N\,(t + t_o) = N\,(t_o)\,F(t). \qquad (2.40)$$

It is interesting that the only functions *F(t)* which satisfy (2.40) for all real values of the elapsed time t and the starting time t_o are exponential functions. This may be seen from two special instances of equation (2.40), namely, $N\,(t_o) = N\,(0)\,F(t_o)$ and $N\,(t + t_o) = N(0)F(t + t_o)$, from which

$$
\begin{aligned}
N\,(t + t_o) &= N\,(0)\,F(t + t_o) \\
&= N\,(t_o)\,F(t) \qquad (2.41) \\
&= N\,(0)\,F(t_o)\,F(t).
\end{aligned}
$$

---

[6]Warning: there are two different uses of the adjective *linear*. The expression is said to be linear in x because its highest power of x is the first, and when plotted as a function of x, the expression yields a straight line. On this basis, most physical scientists call the relation (2.38) linear for any constants a and b. However, a function $f\,(x)$ is often said to be linear in x only if for any scalar constant λ, $f\,(\lambda x) = \lambda f\,(x)$, and on this basis most mathematicians would call (2.38) linear only if $a = 0$.

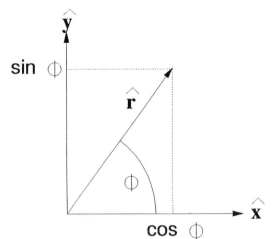

Figure 2.1. Sine and cosine functions can be defined as components of a unit vector in two dimensions.

In other words, for all t and t_o,

$$F\,(t + t_o) = F(t)\,F(t_o). \qquad (2.42)$$

Only exponential dependence behaves this way:

$$F(t) = A^t \equiv e^{-\Gamma t} \equiv \exp(-\Gamma t), \qquad (2.43)$$

where evidently the constant $A = F(1) = \exp(-\Gamma)$ and $e \equiv \exp(1)$ is the real number $e = 2.7182818\ldots$. Where does this number e come from? After all, we could equally well express $F(t)$ by a power proportional to t of any positive, real constant, for example by $10^{-\alpha\,t}$ or by $2^{-\beta\,t}$ where $A = 10^{-\alpha} = 2^{-\beta}$. We'll see what's special about e below (Section 2.9) when we look at derivatives. For *decay* processes, $\Gamma > 0$ and $0 < A < 1$.

Two other common functions are the cosine and sine functions, $\cos(\phi)$, $\sin(\phi)$. Unlike the exponential function, the sine and cosine are *periodic* functions. It is simplest to think of these functions as the x and y components of a unit vector $\hat{\mathbf{r}}$ which is rotated counter-clockwise in the xy plane by an angle ϕ from the x-axis:

$$\begin{aligned}\cos(\phi) &= \hat{\mathbf{r}} \cdot \hat{\mathbf{x}} \\ \sin(\phi) &= \hat{\mathbf{r}} \cdot \hat{\mathbf{y}}\end{aligned} \qquad (2.44)$$

where $\hat{x}$, $\hat{y}$ are unit vectors along the x- and y-axes, respectively (see Fig. 2.1). The domains of the cosine and sine functions are the real numbers $\phi \in \mathbb{R}$, and their ranges are given by the interval $[-1, 1] \in \mathbb{R}$. From Fig. 2.1 and the fact that $\hat{r}$ points along the y-axis if its angle $\phi = \pi/2$, one sees that

$$\cos(\phi) = \sin(\phi + \pi/2). \tag{2.45}$$

Of course, many other functions are possible, including the power-law relation

$$g(x) = A \, x^p \tag{2.46}$$

where the coefficient A and the power p are real constants and for p nonintegral, $x \geq 0$. A table of common functions and constants in Maple is found in Appendix C.

Functions can also be combined by addition, multiplication, and composition to create new functions. Thus from the two functions

$$f : x \mapsto f(x), \; g : x \mapsto g(x), \quad x \in \mathbb{R} \tag{2.47}$$

we can form

$$\begin{aligned} (f + g) : x & \mapsto (f + g)(x) & \equiv f(x) + g(x) \\ (fg) : x & \mapsto (fg)(x) & \equiv f(x) \, g(x) \\ f \circ g : x & \mapsto (f \circ g)(x) & \equiv f(g(x)). \end{aligned} \tag{2.48}$$

If $f(x)$ is the linear function $f : x \mapsto a + bx$ and $g(x)$ is the power-law function $g : x \mapsto c \, x^2$, their sum is the quadratic relation

$$(f + g)(x) = a + bx + cx^2 , \tag{2.49}$$

and their product is

$$(fg)(x) = cx^2(a + bx), \tag{2.50}$$

whereas their compositions are[7]

$$\begin{aligned} (f \circ g)(x) &= \; a + bcx^2 \\ (g \circ f)(x) &= \; c(a + bx)^2. \end{aligned} \tag{2.51}$$

---

[7]Recall that we are considering only simple functions which map arbitrary real arguments onto real values. When considering the product and composition of more general functions, one must ensure that the relevant domains and ranges are of the right types and values so that the product and composition of the functions make sense.

Note that the composition of functions is not generally commutative $(f \circ g \neq g \circ f)$. However, it is associative: $f \circ (g \circ h) = (f \circ g) \circ h$. (Why?)

If the composition of f with g gives the identity function, that is,

$$(f \circ g)(x) = x, \tag{2.52}$$

then f and g are *functional inverses* of each other. Thus $\ln(x)$, $x > 0$, is the functional inverse of $\exp(x)$, and $\arcsin(x)$ is the functional inverse of $\sin(x)$, $-\pi/2 \leq x < \pi/2$.

Another way to generate new functions from old ones is to take the rate of change of the function value with respect to its argument. This may be recognized as the traditional definition of a *derivative:*

$$\frac{df(x)}{dx} = \lim_{\delta \to 0} \frac{f(x+\delta) - f(x)}{\delta}. \tag{2.53}$$

You probably used this definition in calculus class to determine the derivatives of some functions. The power-law function $f(x) = cx^p$, $p \neq 0$, for example, has the derivative

$$
\begin{aligned}
\frac{df(x)}{dx} &= \lim_{\delta \to 0} c \frac{(x+\delta)^p - x^p}{\delta} \\
&= \lim_{\delta \to 0} c \frac{x^p + p\delta x^{(p-1)} + \dots - x^p}{\delta} \\
&= p(cx^{(p-1)}).
\end{aligned} \tag{2.54}
$$

2.8   *Plotting

Plots are valuable for visualizing functions, but since manual plotting can be tedious, we often avoid plotting relations simply to save time. With Maple, the computer handles the tedium, and we can and should make abundant use of the plotting utility. The Maple commands are simple: to plot the sine function, the command is

$$>\texttt{plot(sin);} \tag{2.55}$$

The process is similar for the exponential function :

$$>\texttt{plot(exp)};\tag{2.56}$$

The plot of the sine function looks reasonable, but the exponential function? It is close to zero over most of the plotted domain. Maple uses a *default* domain of $-10..10$ (which is Maple's way of writing "from -10 to 10") if the domain is not specified, and $\exp(10)$ is so large and growing so rapidly that most values of $\exp(x)$ at lower x are negligible. To get a clearer image of the functional dependence of the exponential function at small x, we need to specify a smaller domain, which can be added as a second argument:

$$>\texttt{plot(exp, 0..3)};\tag{2.57}$$

gives the plot for the domain $(0, 3)$. Compare this to the function over the domain $(-2, 1)$:

$$>\texttt{plot(exp, -2..1)};\tag{2.58}$$

Inspection shows a curious scaling property: the shape of the function for $x \in (0, 3)$ is the same as for $x \in (-2, 1)$, only the magnitude of the function is different. An analytic statement of this scaling property is

$$\exp(x)\exp(2) = \exp(x+2), \quad x \in (-2, 1)\tag{2.59}$$

and this is just a special case of (2.42). The scaling property is thus a necessary consequence of demanding that the factors (i.e., the exponential function) of fractional decay (or growth) depend only on the elapsed time and not on the initial time. To return from the plot to the Maple worksheet, press <Alt>-F, X or double-click on the plot control button at the upper right of the plot window. If you want to save the plot, you can reduce it to an icon by clicking on the down triangle at the upper left-hand corner of the plot window. Since for all real x, $\exp(x) > 0$, condition (2.42) shows that $\exp(x)$ is a monotonically increasing function of x; that is, if $x_2 > x_1$, then $\exp(x_2) > \exp(x_1)$.

Monotonically increasing functions describe one-to-one maps: every x value is uniquely mapped to a functional value $f(x)$; no other x gives the same functional value. Such functions have inverses f^{-1} which describe the inverse mapping operation $f^{-1} : f(x) \mapsto x$. A plot of f^{-1} is related to one of f by a reflection in the line which bisects the axes. For example, compare the function exp and its inverse ln in a single plot, with both domain and range equal $-5..5$:

$$> \texttt{plot}(\{\exp, \ln\}, -5..5, -5..5); \qquad (2.60)$$

Can you see the relation?

Note that both exp and $\exp(x)$ are referred to as *functions*. Maple needs to be more precise and calls exp a functional *operator* and $\exp(x)$ a functional *expression*. The above plots were of functional operators, but we can also plot expressions:

$$> \texttt{plot}(\exp(x), x = -1.5..1.5) \qquad (2.61)$$

and this also serves to label the abscissa (horizontal axis).

Now plot the cosine and sine functions for the domain $(-2\pi, 2\pi)$. Verify the relation (2.45), which states that both functions have the same shape but that the cosine curve is a quarter cycle or $\pi/2$ radians behind the sine curve. Next, try plotting the tangent function (tan) over the same domain. You'll see an obvious problem with the automatic scaling of the ordinate (vertical axis) because the function is unbounded over this domain. You can limit the plotted range by adding a third parameter:

$$> \texttt{plot(tan,-2*Pi..2*Pi,-5..5);} \qquad (2.62)$$

The Maple symbol for composition of functions is @ . Thus

$$(\texttt{ln @ exp})(\texttt{x}) \;=\; \texttt{x} \;. \qquad (2.63)$$

Try plotting $(\sin \circ \exp)(x)$ and compare the plot to that of $\sin(x) * \exp(x)$ over the same domain. Finally, plot both functions together with the command

$$> \texttt{plot}(\{\sin \texttt{ @ } \exp, \sin * \exp\}, -2..2); \qquad (2.64)$$

The curly brackets {}, by the way, are Maple's symbol for a set. The above plot command thus asks Maple to plot both of the functions in the set.

2.9   *diff and D

Let's return once more to the example of Section 2.5 on radioactive decay in order to learn about Maple's differentiation operators, diff and D.

A useful approach to finding the derivative of $F(t) = \exp(-\Gamma t)$ is to take the limit in which t_o in (2.42) is a vanishingly small time increment denoted by the infinitesimal dt. The infinitesimal change in $F(t)$ from t to $t + dt$ is then

$$dF(t) = F(t + dt) - F(t) = F(t)[F(dt) - 1]. \qquad (2.65)$$

Now an expansion of $F(t)$ in a power series in dt gives

$$F(dt) = 1 + f_1 dt + ... \qquad (2.66)$$

where we can ignore higher-order terms in the infinitesimal dt, and the constant coefficient f_1 must be negative since $F(t)$ decreases with increasing elapsed time. When (2.66) is substituted into (2.65), it implies that $F(t)$ is proportional to its first derivative:

$$\frac{dF(t)}{dt} = f_1 F(t) \qquad (2.67)$$

with the initial condition $F(0) = 1$. Let's have Maple check this out. Try redefining the function $F(t)$ as in (2.20) and entering

$$> \text{diff}(F(t), t); \qquad (2.68)$$

which asks Maple to calculate the first derivative of the functional expression. The expression diff(F(t), t) is one Maple way of writing the derivative $dF(t)/dt$. Another way is to use the functional operator with D: D(F). Try calculating that, too. Are D(F)(t) and diff(F(t),t) equal? Note that the Maple derivative operator D operates only on functional operators to give a new operator, whereas diff acts on functional expressions to give new expressions. Maple's response for the derivative should allow you to relate the constant f_1 to Γ. Because

$$\Gamma = -f_1 = -\frac{1}{F(t)} \frac{dF(t)}{dt}, \qquad (2.69)$$

it may be called the *(fractional) decay rate.*

In fact, exponential decay results whenever the fractional rate of decay is constant, independent of how many nuclei have already decayed. This is the case when each excited nucleus has the same probability, namely Γdt, of decaying in the next time interval dt, independent both of what the other nuclei have done and of what its own past history has been. In fact, the nuclei are well separated and have no significant interaction with each other, and the state of each nucleus completely determines its properties, including its fractional decay rate.

Of course, the same differential equation (2.67) describes *growth* if $f_1 > 0$, and Maple's differentiation (2.68) confirms that such growth is exponential. The growth of various populations, in particular of cancer cells, approximates this behavior because the fractional growth rate is nearly constant and independent of population size. Exponential growth is, of course, just the time reversal of exponential decay, and just as decay could be characterized by a half-life, so exponential growth can be characterized by a doubling time t_double, related to the fractional growth rate f_1 by

$$f_1 = \frac{\ln(2)}{\text{t\_double}}. \tag{2.70}$$

On a plot of the functional dependence of any simple function `f(x)`, the first derivative `D(f)(x) = diff(f(x),x)` is the slope of the curve. Check this visually by having Maple plot both `f(x)` and its derivative `D(f)(x)` for several simple functions, such as

```
>   f := x -> x^2;
>   f := x -> sin(x^2);          (2.71)
>   f := x -> exp(-x^2);
```

over domains $x \in (-5, 5)$ or so. The plots of the function can be distinguished from that of its derivative in the above cases because all the functions are even: $f(x) = f(-x)$. You can simplify your typing a bit by using the symbols ", " ", and " " " for "the last expression," the "next to last expression," and "the second from last expression," respectively. Thus, if the function definition was your last expression, the command

```
>   plot({",D(")},x = -5..5);          (2.72)
```

should work. Another possibility to save typing is to scroll back on the worksheet screen to a previous command which you can use. Simply highlight it and press <Ctrl>-C to copy it to the clipboard. Then move the cursor to the desired insertion point and press <Ctrl>-V to copy it from the clipboard to the worksheet. If the command needs changes, simply edit it. When it's correct, simply press <Enter> with the cursor located somewhere on the command, and it will be executed.

Repeat the plots of functions and their derivatives for the functions $f(x) = c^x$ with $x \in (-2, 2)$ where the constant c is 2, 3, or 4. You can identify the function in each case because $f(0) = 1$. Note that in all cases, the derivative (and hence the slope) of the function is proportional to the function itself, and that when $c = 3$, the slope is almost equal to the function. Finally check graphically that when $c = e$ (Maple's symbol for the constant $e = \exp(1)$ is E), the slope is, in fact, equal to the function. This result provides the answer to a question posed above, namely what is special about the value of e. We see now that e is the value for c which makes c^x equal to its slope and hence to its derivative.

> **Exercise 2.4.** The n-th derivative of a function $f(x)$ can be expressed in Maple by either `diff(f(x),x$n)` or `(D@@n)(f)(x)`. Show with Maple that the fourth derivatives of the three functions $\exp(x)$, $\sin(x)$, and $\cos(x)$ are equal to the functions themselves.

To finish this chapter, compare plots of $\sin(x)$ with its slope. You knew it already, of course, but it's nice to see visually that the derivative $D(\sin)(x) \equiv d\sin(x)/dx$ has the same periodic, sinusoidal shape as $\sin(x)$ but is displaced by a quarter of a cycle in x: $D(\sin)(x) = \cos(x)$.

2.10   *Problems

1. The following expressions are ambiguous; they may be calculated in at least two different ways, with different results depending on the order of operations. Use parentheses to indicate different interpretations of the expressions and give the results. Finally, run the ambiguous expressions on Maple. Formulate a general

rule for the way Maple handles expressions with several operators
of the same priority level.

(a) $4 - 3 + 2$

(b) $4/2/2$

(c) $8/4 * 2/4$

2. Before you have Maple compute them, *estimate* the values of
$3\hat{}(4\hat{}5)$ and $(3\hat{}4)\hat{}5$. Explain why one should have about 10
digits whereas the other should have roughly 500. Then check
your reasoning with Maple. To keep these numbers "in perspec-
tive," remember the total number of protons in the universe is
"only" about 10^{80} (see problem 4 of Chapter 1).

3. The factorial function $n!$ is given for any positive integer by the
iterative definition
$$0! \equiv 1$$
$$n! = n * (n - 1)! \tag{2.73}$$

For example, $1! = 1$, $2! = 2*1$, $3! = 3*2*1 = 6$, $4! = 4*3*2*1 = 24, ...$ Use Maple to compute 20! (enter it as it is written: a
number followed by an exclamation point: 20!). Next try the
binomial coefficient
$$\frac{10!}{3! * 7!} \equiv \binom{10}{3}. \tag{2.74}$$

After you have found it directly, try to calculate the Maple function binomial(10,3) to see if you get the same result.

4. Express the functions below in terms of sums, products, and
compositions of the basic functions $f(x) = 3x$, $g(x) = x^2$, and
$h(x) = \exp(x)$.

(a) $3x + x^2$

(b) $\exp(3x^2)$

(c) $3\exp(x + 2x^2)$

5. Check derivatives of combinations of functions:

(a) Find the derivatives of the functions in parts (a) to (c) of problem 4.

(b) Write down general rules for finding the derivatives of sums, products, and compositions of two functions. Verify your results by letting Maple evaluate

$$> \texttt{D(f + g); D(f * g); D(f@g);} \qquad (2.75)$$

6. The change s in height of a ball (ignoring air resistance) thrown with an initial vertical component of velocity v_0 is well known to be

$$s = v_0 t - \frac{1}{2}gt^2$$

where $g = 9.8$ m/s^2. Plot together the heights as a function of time for four trajectories, with $v_0 = \pm 5, 0$, and 10 m/s, with t in the domain $(0..2.5)$ seconds and s in the range $(-6..6)$ meters. Would it be possible to throw two balls with different velocities, starting from the same position at the same time, which would later collide? Explain why the trajectories in space should have the same shape (to within a scaling factor for the horizontal direction) for any initial horizontal component of the velocity.

7. Explore the help files in Maple, especially the examples, to find a compact way, using the seq call, of defining as a set the family of functions

$$\sin(nx)/x, \quad n = 1, 2, \ldots, 10, \qquad (2.76)$$

and write the Maple plot command to combine plots of all ten functions over the domain $x \in (-4, 4)$. (Check to make sure your command really works!)

8. Use Maple and the reference data on chemical isotopes in the directory 'atom' (see Chapter 1) to find at what time in the past the abundances of the isotopes U-235 and U-238 would have been equal and how their abundances at that time would have compared to the present total abundance of uranium on earth. Assuming that the isotopes were formed in equal amounts in a supernova explosion, this information will tell you when that explosion

occurred and how much larger the abundances were at that time than today, relative to the stable elements which form the surface of the earth. [Hint: In Maple, assign a set of two equations to be solved

> eqs:={U235=A*2^(t/T235),...};

where U235 is the present abundance of U-235, T235 is its half-life, and A and t are the isotope abundance and the time (with respect to the present) when the two abundances were equal. Express the half-lives in units of gigayears to keep Maple from choking on large integers. Then use the Maple command

> solve(eqs,{A,t});

to find the unknown variables A, t.]

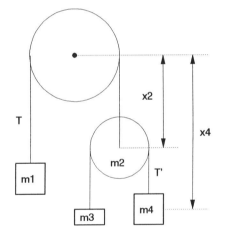

Figure 2.2. The double Atwood machine. The larger pulley is fixed, but the smaller one, of mass m2, can move; both are frictionless and have negligible moments of inertia. The gravitational acceleration is g.

9. Consider the double Atwood machine in Fig. 2.2. Assume that the pulleys are frictionless and have negligible moments of inertia, and that the strings are of fixed length. From Newton's second law applied to each isolated mass, write down the acceleration of

each mass in terms of all the forces acting on that mass. (The only forces acting in this idealized problem are the gravitational forces and the tensions.) Relate the acceleration of $m1$ to that of $m2$, and the acceleration of $m2$ to the accelerations of $m3$ and $m4$. (Hint: it may be easier to relate the heights of the masses and differentiate with respect to time.) Construct a set of four equations for the two tensions and the two accelerations $a2$ and $a4$, and use this set of equations as well as a set of unknown variables in Maple's `solve` command to find the acceleration of mass $m4$. Write down both the Maple commands needed and the final result. Note that Maple finds it convenient to define a term it labels %1.

2.11   Chapter Summary

2.11.1   Concepts

assignment statements

tokens

parsing

strings

additive and multiplicative inverses

functions

domain, range

linear, quadratic and cubic functional dependence

exponential decay and growth

periodic functions

functional composition

inverse functions

differentiation

commutative and associative binary operations

the factorial function and binomial coefficient

2.11.2 Maple Commands

```
^  *  /  +  -
```

```
:=   :  ;  ,,  ,,,,  ,,,,,,
```

```
= <>  <  >  <= >=
```

```
escape characters ?  !  #  \
```

```
help
```

```
example
```

```
quit, done, stop
```

```
evalf
```

```
simplify
```

```
Digits
```

```
solve
```

```
sqrt, sin, cos, exp, ln
```

```
plot
```

```
solve
```

```
diff, D
```

```
binomial
```

```
seq
```

Chapter 3

Approximations of Real Functions

Often it's convenient or even necessary to approximate the value of a function at a point or over a small domain of points. Approximations can, of course, be very precise at times, and we often use them as though they were exact. Whenever you compute $\sin(0.1)$ on a calculator or in Maple, for example, the value must be approximated. A pocket calculator approximates the value to a limited accuracy of usually less than 12 decimal places, but Maple can approximate such numbers to what is for most practical purposes essentially arbitrary accuracy.

Approximations are common in the physical sciences. For example, in order to determine the motion of projectiles near the surface of the earth, we usually approximate the true gravitational field by a uniform one with a constant value of acceleration; to predict the motion of a pendulum, we approximate the functional dependence of the restoring force on the displacement. Linear approximations are the most common, but higher-order power-series and asymptotic series also play an important role.

Maple's ability to compute, differentiate, and plot simplifies evaluations of the accuracy and suitability of various approximations. In addition, Maple's talents at generating series expansions and calculating asymptotic forms can help to derive various approximations.

In this chapter, we begin by investigating linear approximations. Their applications to simple-harmonic oscillators warrants a separate section. Higher-order approximations, especially Taylor series expansions, are considered next. Even and odd functions, products of power series, asymptotic series, and limits are treated as well.

Work with Maple gives more experience with the differential operator D, which is useful in developing Taylor series expansions. The commands plot, series, asympt, limit, map, assign, unassign, unapply, expand, sum, nops, op, degree, and select are also exercised (some, appropriately enough, in the exercises). Finally, the concept of sequences, sets, and lists in Maple are introduced.

Most of the material in Sections 3.1–3.6 is important preparation for the rest of this text. In particular, power-series expansions (Sections 1 and 3) are needed in Chapter 6, the simple harmonic oscillator (Section 2) is extended in Chapters 7 and 8, and the discussion of power series (Section 3) and even and odd functions (Section 4) is needed in Chapter 9.

3.1   *Linear Approximations

The simplest approximation of $\exp(x)$ for small x is $\exp(0) = 1$. Recall the plot of the exponential function from chapter 2: for a given value of x on the horizontal axis, the plotted curve directly above that point has a height equal to the value of the function, and the approximation $\exp(x) \approx \exp(0) = 1$ represents the exponential curve by the same value at every x; its plot is a horizontal line. The approximation is exact at one value of x, but it totally ignores any dependence of the function on its argument. A far better *linear* (straight-line) *approximation* at small x is the one which not only touches the curve $\exp(x)$ at $x = 0$ but also has the same slope as the curve at that point. Recall that the slope of a curve $f(x)$ is the first derivative

$$\frac{df(x)}{dx} \equiv D(f)(x) \tag{3.1}$$

where the last two forms correspond closely to convenient Maple expressions. Generally, for any continuous function $f := x \to f(x)$, the last form treats the differential $D(f)$ as a new functional operator on x: $D(f) : x \to D(f)(x) \equiv df(x)/dx$. The slope of the curve $f(x)$ at x_0

can thus be written

$$D(f)(x_0) \equiv \left[\frac{df(x)}{dx}\right]_{x=x_0} \tag{3.2}$$

As you remember from the last chapter, the slope of the exponential curve is the exponential curve itself:

$$D(\exp) = \exp . \tag{3.3}$$

In particular, the slope of the curve at $x = 1$ is $\exp(0) = 1$. Now any function $y(x)$ whose curve is a straight line has the form $y(x) = a + bx$ and can thus be defined in Maple by

$$y := x \rightarrow a + bx \tag{3.4}$$

where a and b are constants. To find the best linear approximation of $\exp(x)$ near $x = 0$, we must choose the constants so that

$$\begin{aligned} y(0) &= \exp(0) \\ D(y)(0) &= D(\exp)(0) . \end{aligned} \tag{3.5}$$

These two conditions immediately give $a = b = 1$. The best linear approximation of $\exp(x)$ near $x = 0$ is thus

$$\exp(x) \approx 1 + x . \tag{3.6}$$

The approximation is exact only in an infinitesimally small neighborhood of $x = 0$. Deviations of $\exp(x)$ from $1 + x$ at small x vary as x^2. One writes

$$\exp(x) = 1 + x + O(x^2) , \tag{3.7}$$

which means that in the limit of vanishingly small x , the difference $\exp(x) - (1 + x)$ vanishes as a constant times x^2. One says that $\exp(x)$ equals
$1 + x$ *plus terms of order* x^2 (or, equivalently, *terms of second order* in x).

Exercise 3.1. Plot $1 + x$ and $\exp(x)$ together over the domain $-1 < x < 1$ to compare the exponential function with its linear approximation. (Review instructions for plotting in the last chapter, if you need help.)

Exercise 3.2. Try the Maple command
```
> for x from -1.  to 1 by .1
do x,exp(x),1+x,1-(1+x)/exp(x) od;
```
This should make a table of values to compare the function $\exp(x)$ with its linear approximation at $x = 0$. (The command is fairly self-explanatory, but note the decimal after the -1, which makes the number approximate instead of exact, and note the *sequence* of values to be tabulated. That is a simple way to get everything in the sequence printed on a single line. Also note the form of the
```
for...from...to...by...do...od;
```
command. It is often useful. More explanation and examples of its use can be found in the Help menu.) Modify the table to print out the list for $-0.16 < x < 0.16$ in steps of 0.02.

It's easy to get Maple to make a table of values showing $\exp(x)$, $1 + x$, and the relative error $1 - (1 + x)/\exp(x)$ at different values of x. (See Exercise 3.2.) You can see that as long as the magnitude of x is less than about 0.14, the function $\exp(x)$ and its linear approximation at $1 + x$ agree within 1%. However, the relative error becomes worse as $|x|$ increases, and at $x = 1$ it exceeds 26% whereas at $x = -1$, it reaches 100%. If we need to approximate values of the $\exp(x)$ for x in a domain close to some value other than $x = 0$, say x_0, a better linear approximation would be the one whose value and slope match that of $\exp(x)$ at x_0, namely

$$y(x_0) = a + bx_0 = \exp(x_0)$$
$$D(y)(x_0) = b = \exp(x_0). \tag{3.8}$$

The best linear approximation at x_0 is thus

$$\exp(x) = \exp(x_0) + \exp(x_0)(x - x_0) + O\left[(x - x_0)^2\right]. \tag{3.9}$$

This approximation could also have been obtained from the linear approximation at $x = 0$ by making the substitution $x \to x - x_0$. We recognize that when terms of order n in some variable are multiplied by a constant, they are still terms of order n in that variable.

These ideas are easily extended to arbitrary continuous functions $f(x)$:

$$f(x) = f(x_0) + D(f)(x_0)(x - x_0) + O\left[(x - x_0)^2\right] \qquad (3.10)$$

is the best straight line approximation to a function $f(x)$ in a small neighborhood of x_0.

As an example, let's estimate how the gravitational field of the earth varies as a function of height close to sea level. For the purpose of this calculation, we will approximate the earth by a sphere of uniform density ρ and radius R. The gravitational field gives the acceleration of a small, freely falling "test" body. As we saw in Chapter 1, at the earth's surface the field has a magnitude

$$g(0) = Gm/R^2, \qquad (3.11)$$

which is the same as if the entire mass m of the earth were concentrated at its center, a distance R away. At a height y above sea level, the magnitude of the gravitational field is

$$g(y) = Gm(r + y)^{-2}, \quad y \geq 0 \qquad (3.12)$$

The linear approximation of $g(y)$ at $y = 0$ is

$$\begin{aligned} g(y) &\approx g(0) + yD(g)(0) \\ &\approx g(0)(1 - 2y/R). \end{aligned} \qquad (3.13)$$

Thus, the gravitational field decreases by about 1% for every increase in height y by $R/200 \approx 32$ km. What if we dig a tunnel and go below sea level? Does the gravitational field increase? No, it *decreases,* because that part of the earth at larger distances from its center no longer exerts any net pull on us.

One way to see this is to imagine gravitational field lines. The total number of field lines is proportional to the total mass, and the field strength at any position is the flux density, that is, the number of lines passing through a unit area oriented perpendicular to the lines. Field lines are continuous except where they begin, at their sources, and can never cross since their density at the crossing point would be infinite. The spherical symmetry of the problem dictates that the field lines, themselves, must be radial. In fact, the field lines from any spherical shell of mass must radiate outward from the shell; none can radiate inward since such lines would have to cross others. Therefore, the field strength at any point outside the shell is the same as if the entire mass of the shell were concentrated at its center, and inside the shell the field vanishes.

The field of the earth, modeled by a solid sphere of radius R, is simply the sum of contributions from all the spherical shells which constitute the solid. The field at a radius $r = R + y$ is the same as the field due to a point mass at the center whose effective mass is the mass of that part of the earth inside the spherical volume of radius $R + y$. For $y > 0$ the effective mass is the total mass m of the earth, but below sea level, where for $y < 0$, only mass at distances less than $R + y$ contributes to the field. Thus, the effective mass is less, namely $m(1 + y/R)^3$, in our model of an earth with constant mass density, and the gravitational field is

$$g(y) = Gm(R + y)/R^3 = g(0)(1 + y/R), \quad y \le 0. \qquad (3.14)$$

You would therefore have to travel twice as far below sea level as above (but still with $\mid y \mid \ll R$) to obtain the same reduction in the gravitational field.

3.2   Simple Harmonic Motion

One of the simplest and oldest man-made oscillating systems is the mass on a light-weight string: the pendulum. Galileo (1564–1642) is credited with noting the important property in lanterns hung in a church: their period of swing, the time to complete one cycle, was nearly independent of the amplitude of motion. Pendula could, and of course did, serve to keep the pulse of clocks for the next three centuries.

We consider here the motion of a pendulum moving in a vertical plane (see Fig. 3.1). The mass m on such a pendulum follows a path

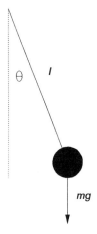

Figure 3.1. The pendulum. Its motion is simple harmonic if the linear approximation $\sin(\theta) \approx \theta$ can be used in describing the restoring force.

along the fixed arc of a circle whose radius is equal to the length l of the string. The tension in the string is sufficient to balance the component of the gravitational force along the string and to provide the necessary centripetal force to keep the mass moving along the arc. The distance of the mass along the arc from its equilibrium position is given by $s = l\theta$, where θ is the angle made by the string to the vertical. The acceleration of m along the arc is due solely to the component $-mg\sin\theta$ of the gravitational force perpendicular to the string. The minus sign indicates that the component of the gravitational force acts as a *restoring force*: it pulls a displaced pendulum back toward its equilibrium position. Newton's second law provides the *equation of motion*:

$$m\frac{d^2(l\theta)}{dt^2} = -mg\sin(\theta). \tag{3.15}$$

Physically, the oscillations occur because of the combined actions of the restoring force and the inertia of the mass. The restoring force increases with increasing displacement of the mass from its equilibrium position. The force continues to accelerate the mass toward the origin ($\theta = 0$) but falls to zero as the origin is reached. By then, however, the pendulum has acquired enough momentum to swing past the origin to the other side, and it continues in the same direction until the restoring force finally turns it around.

An analytic solution for the pendulum is fairly simple if the function is approximated by a linear function:

$$\sin(\theta) = \sin(0) + \theta D(\sin)(0) + O(\theta^2) \approx \theta. \tag{3.16}$$

The equation of motion, in terms of the distance $s = l\theta$ along the arc becomes an equation for *simple harmonic motion*:

$$\frac{d^2 s}{dt^2} = -\frac{g}{l} s. \tag{3.17}$$

We already know the solution of this equation. Recall that in uniform circular motion, the acceleration has a constant magnitude and is directed in the direction opposite to the displacement of the object from the center of the orbit. If the vector $\mathbf{r}$ is the position of the object in uniform circular motion, we can thus write

$$\frac{d^2 \mathbf{r}}{dt^2} = -\omega^2 \mathbf{r}, \tag{3.18}$$

where the constant angular velocity ω is the factor that relates both the speed to the radius and the magnitude of the acceleration to the speed. If we take the orbit of the motion to lie in the xy plane, the position vector can be written in components as

$$\mathbf{r} = x\hat{\mathbf{x}} + y\hat{\mathbf{y}}, \tag{3.19}$$

and the acceleration becomes

$$\frac{d^2 \mathbf{r}}{dt^2} = \frac{d^2 x}{dt^2}\hat{\mathbf{x}} + \frac{d^2 y}{dt^2}\hat{\mathbf{y}}. \tag{3.20}$$

Both components of the motion obey an equation of the form

$$\frac{d^2 x}{dt^2} = -\omega^2 x, \tag{3.21}$$

which is identical to that for the linearized pendulum.

We can conclude that the motion of an object with a linear restoring force is just the projection of uniform circular motion, and further, the constant of proportionality giving the magnitude of the acceleration per unit displacement is exactly the square of the angular frequency. In the case of the pendulum,

$$\omega = \frac{2\pi}{T} = \sqrt{\frac{g}{l}}\,, \tag{3.22}$$

where T is the period, that is, the time to complete one full cycle. Since the projection of uniform circular motion along a given direction can be written

$$x = x_0 \cos\phi\,, \quad \phi = \phi_0 + \omega t\,, \tag{3.23}$$

where x_0 and ϕ_0 are constants given by the initial conditions and representing the amplitude of motion and the initial (phase) angle, the solution for the position s of the pendulum must be analogous:

$$s(t) = s_0 \cos(\phi_0 + \omega t)\,. \tag{3.24}$$

Straightforward differentiation confirms that this is indeed a solution to the linear equation of motion (3.17).

The simple analysis above can also serve to show why simple harmonic motion is so common. Such motion can occur practically whenever a system has a point of stable equilibrium. By definition, a system at rest at a point of equilibrium will stay there; the net force on the object vanishes at any equilibrium point. The equilibrium is *stable* if the system returns to equilibrium whenever it is displaced by sufficiently small amounts. For a one-dimensional system, this means that the force on the system is negative for small positive displacements and positive for small negative ones. Most physical forces are smooth, so it is likely that the restoring force can be well represented by a linear approximation for sufficiently small displacements. If x measures the displacement from equilibrium, then the linear approximation to the force must have the form $F = -kx$ where the magnitude k of the slope gives the "stiffness" of the restoring force. If damping can be ignored, the equation of motion is then

$$\frac{d^2 x}{dt^2} = -\frac{k}{m} x \, , \tag{3.25}$$

whose solution, as above, is simple harmonic motion with an angular frequency $\sqrt{k/m}$.

> **Exercise 3.3.** In one-dimensional conservative systems, the force $F(x)$ can be represented by minus the slope of a *potential $U(x)$*:
>
> $$F(x) = -\frac{dU(x)}{dx} \tag{3.26}$$
>
> Show that according to the above discussion, the potential must have a minimum at any point of stable equilibrium, and that the linear approximation of the force corresponds to a parabolic approximation of the potential minimum. Find the general form of the potential U which gives the linear force $F = -kx$.

3.3   Power-Series Expansions

As useful as the linear approximation is, we often need more accuracy. The linear expansion of the exponential function $\exp(x)$ about $x = 0$, for example, gave large errors when used to compute $\exp(1)$ or $\exp(-1)$. Even in those cases where we must be satisfied with a first-order approximation, we often need to advance to the next step in order to estimate the error of our approximation. The next step is to approximate a function by a curve whose slope is changing at a constant rate. The rate of change of the slope is, of course, the second derivative of a function, and a function quadratic in the displacement has three coefficients which can be chosen to match the function and its first and second derivatives at one point.

For example, the best quadratic or second-order approximation to $\exp(x)$ in an infinitesimal neighborhood of $x = 0$ has the form

$$y := x \rightarrow a + bx + cx^2 \, , \tag{3.27}$$

and the matching conditions

$$\begin{aligned}
y(0) &= \quad a = \exp(0) = 1 \\
D(y)(0) &= \quad b = \exp(0) = 1 \\
D^2(y)(0) &= \quad 2c = \exp(0) = 1
\end{aligned} \qquad (3.28)$$

give the values $a = b = 2c = 1$. The symbol $D^2(y)$ represents the second derivative of the function y, but it is more accurate to think of D^2 as the composition $D \circ D$ rather than a multiplication, since $D^2(y) = D(D(y))$. Note that the coefficients a, b have the same value as in the linear approximation. Indeed, the quadratic term does not enter the matching conditions until the second derivatives of $y(x)$ and $\exp(x)$ are equated. Thus the second-order approximation for $\exp(x)$ correct up to but not including third order, is

$$\exp(x) = 1 + x + \frac{1}{2}x^2 + O(x^3). \qquad (3.29)$$

If we now compare the percent errors at $x = \pm 1$, we find that our error is down from 26% to 8% at $x = 1$ and down from 100% to 36% at $x = -1$. That marks a definite improvement, but clearly we could do better. Continuing to higher powers of x, one obtains

$$\exp(x) = \sum_{k=0}^{n-1} \frac{x^k}{k!} + O(x^n). \qquad (3.30)$$

This series can be shown to converge in the limit $n \to \infty$ for any finite x. It can be used to approximate the constant e : up to but not including 7th order, one obtains $e \approx 2.718056$, which is correct to four places. The error is the order of the first neglected term, in this case $1/7!$, which is roughly 2×10^{-4}.

The generalization of such an expansion to an arbitrary function $f(x)$ about the point x_0 gives a power series known as the *Taylor* expansion

$$f(x) = \sum_{n=0}^{\infty} \frac{D^n(f)(x_0)}{n!}(x - x_0)^n. \qquad (3.31)$$

This expansion may not converge at all values of x. Where it does converge, $f(x)$ is said to be *analytic*. Any polynomial in x has a finite power-series expansion and is analytic everywhere. Functions like $\ln(x)$

and x^q where q is a not a positive integer or zero have no power series expansions about $x = 0$ but may be expanded about other points, for example about $x = 1$:

$$\ln(x) = \frac{(x-1)}{1} - \frac{(x-1)^2}{2} + \frac{(x-1)^3}{3} - \dots \quad (3.32)$$

Power series for some functions can be found without calculating explicit derivatives. Expressions like $(a+x)^q$, for example, can be expanded as binomials, even when q is not a positive integer or zero:

$$\begin{aligned}(a+x)^q &= a^q + qa^{q-1}x + \frac{q(q-1)}{2!}a^{q-2}x^2 + \dots \\ &= \sum_{n \geq 0} \binom{q}{n} a^{q-n}x^n .\end{aligned} \quad (3.33)$$

It is easily verified that the coefficients are identical to those for a Taylor expansion about $x = 0$. An important special case is

$$\frac{1}{(1-x)} = 1 + x + x^2 + x^3 + \dots, \quad |x| < 1. \quad (3.34)$$

Exercise 3.4. Expand $1/\sqrt{1-x}$ in a power series in x up to but not including terms of order x^3. Compare your quadratic approximation with the exact square root by plotting both over the domain $-.5 < x < .5$.

Exercise 3.5 Expand $1/x$ about $x = 1$. [Hint: a change of variable should make this expansion analogous to the expansion (3.34) above of $1/(1-x)$ in powers of x. Try $x = s + 1$.]

Power series simplify linear operations, since over the domain of convergence of the series, the operations can be applied term by term. Differentiation and integration are such linear operations. Thus if a function $f(x)$ has a power series expansion

$$f(x) = \sum_{n=0}^{\infty} a_n x^n \quad (3.35)$$

which is convergent in a neighborhood of x_0, its slope (first derivative) at x_0 is

$$D(f)(x_0) = \sum_{n=0}^{\infty} n a_n x_0^{n-1} . \tag{3.36}$$

Furthermore, if two convergent power series in x are equal over a neighborhood of $x = 0$, their coefficients must be equal term by term (the double-headed arrow "$\Leftrightarrow$" means "is equivalent to" and hence "iff"):

$$\sum_{n=0}^{\infty} a_n x^n = \sum_{n=0}^{\infty} b_n x^n \Leftrightarrow a_n = b_n, \; n = 0..\infty . \tag{3.37}$$

Exercise 3.6 Prove this relation (3.37) from (3.36) by differentiating n times and setting $x = 0$.

Exercise 3.7 Show that any function $f(x)$ analytic in a neighborhood of x_0 [that is, $f(x)$ has some convergent power series expansion about x_0] is equal to the Taylor series expansion given by (3.31).

3.4 Even and Odd Functions

A function $f(x)$ is said to be *even* iff (if and only if) $f(x) = f(-x)$. It is *odd* iff $f(x) = -f(-x)$. Any function can be written as a sum of even and odd parts:

$$f(x) = f_+(x) + f_-(x)$$

$$f_+(x) := \tfrac{1}{2}[f(x) + f(-x)] \tag{3.38}$$

$$f_-(x) := \tfrac{1}{2}[f(x) - f(-x)] .$$

Obviously even powers of x are even functions, whereas odd powers of x are odd functions. In a power-series expansion of $f(x)$ about $x = 0$, the even part $f_+(x)$ is equal to the sum over even powers, whereas the odd part $f_-(x)$ is given by the sum over odd powers.

More generally, a sum of even functions is even; a sum of odd functions is odd. The product of an even function with an odd one is odd, whereas the product of two even functions or two odd functions is even.

The derivative D of an odd function is even and *vice versa*; D may be considered an odd operator.[1] The composition of an even function with either an even or an odd function is even, but the composition of two odd functions is odd.

The exponential function $\exp(x)$ is neither even nor odd. Its even and odd parts are called the hyperbolic cosine and hyperbolic sine functions, respectively:

$$\cosh(x) = \frac{1}{2}\left[\exp(x) + \exp(-x)\right] = \sum_{n=0}^{\infty} \frac{x^{2n}}{(2n)!} \tag{3.39}$$

$$\sinh(x) = \frac{1}{2}\left[\exp(x) - \exp(-x)\right] = \sum_{n=0}^{\infty} \frac{x^{2n+1}}{(2n+1)!}. \tag{3.40}$$

3.5   Power-Series Products

Often power series must be constructed from products of such series. It is important to include all products of a given power and to omit all terms of any order for which some terms are unknown. This usually means that many products of terms can (and *should*) be dropped. Consider, for example, the product

$$\left[x + 2x^2 + 3x^3 + 4x^4 + O(x^5)\right]\left[x^3 + 3x^4 + O(x^5)\right] \tag{3.41}$$

where $O(x^n)$ means unknown terms of order x^n. The product is

$$x^4 + 5x^5 + O(x^6). \tag{3.42}$$

Because the lowest-order unknown term is of order x^6, it makes no sense to include any explicit terms of that order or higher. Only the first two terms in the first factor could be used.

> **Exercise 3.8.**   Take the square of the power series for $1/\sqrt{1-x}$ (see Exercise 3.4) and show that up to but not

---

[1] By the same token, $D@@3$ and $D@@5$ are also odd operators. In fact, with Maple tokens, many odd operators can be defined.

including third order, the product is equal to $1 + x + x^2$. (Compare the binomial expansion of $(1 - x)^{-1}$ given in Section 3.3.)

Even the product of infinite series can often be put in compact form. As an example, let's show that the product of the series for the exponential function really combines correctly:

$$\exp(x)\exp(y) = \sum_{m=0}^{\infty}\sum_{n=0}^{\infty} \frac{x^m y^n}{m!n!}. \tag{3.43}$$

Consider the first few terms:

$$\exp(x)\exp(y) = 1 + (x + y) + (x^2 + 2xy + y^2)/2! + \cdots \tag{3.44}$$

where we have grouped together terms with $m+n = 0, 1, 2$. The subset of terms with $m + n = N$ is seen to be

$$\sum_{m=0}^{N} \frac{x^m y^{N-m}}{m!\,(N-m)!} = \sum_{m=0}^{N} \binom{N}{m}\frac{x^m y^{N-m}}{N!} = \frac{(x+y)^N}{N!} \tag{3.45}$$

where the binomial expansion (3.33) was recognized. Adding up all such subsets, we find

$$\exp(x)\exp(y) = \sum_{N=0}^{\infty} \frac{(x+y)^N}{N!} = \exp(x+y). \tag{3.46}$$

As useful as Taylor series are, they are generally not the most efficient means of computing functional values. Although for a given number of terms they provide the best expansion in an infinitesimal neighborhood of a point, a different set of coefficients may provide a better approximation over a finite domain. Computers often store functions as expansions in orthogonal polynomials. Ratios of polynomials (known as *Padé approximants*) and continued fractions are other possible approximations. Sometimes it is more convenient to store accurate values of the function at a discrete set of points in the domain. Values at other points can then be found by interpolation. These methods will be discussed further in Chapter 5.

3.6   *Sums, Sequences, Sets, and Lists

To compute power-series expansions, we can use Maple's sum command. For example, try computing the constant from the expansion for $\exp(x)$ about $x = 0$:

$$> \text{sum}(1/n!, n = 0..3);$$
$$> \text{sum}(1/n!, n = 0..6); \hspace{2cm} (3.47)$$
$$> \text{sum}(1/n!, n = 0..12);$$

We can calculate a *sequence* of such approximations with the command

$$> \text{seq}(\text{sum}(1/n!, n = 0..3 * N), N = 1..5); \hspace{1cm} (3.48)$$

Try putting this sequence of numbers into a *list* L by typing

$$> \text{L} := [\text{"}]; \hspace{2cm} (3.49)$$

You can evaluate the list of numbers with the command

$$> \text{evalf}(L); \hspace{2cm} (3.50)$$

To find the percent errors, use the map command:

$$> \text{Lerr} := \text{map}(x- > (1 - x/E) * 100, L); \hspace{1cm} (3.51)$$

which takes each of the five elements L[N], N = 1..5 of the list and computes a new list of values(1 - L[N]/E)*100 where E is $\exp(1)$. Generally, if s is a sequence of expressions, then [s] is a list of those expressions, and {s} is a set. The main difference between a set and a list is that the list is ordered, whereas in a set the order is immaterial, and redundant elements are eliminated. Thus the sequence

$$> \text{s} := 1, 2, \text{a}, 2, 1; \hspace{2cm} (3.52)$$

gives the list [s] = [1,2,a,2,1] and the set {s} = {1,2,a}.

Exercise 3.9. Define a sequence in Maple of values $n * (N - n)/2$ for $n = 1..N$ where $N = 15$. Form the list and the set of this sequence.

3.7 *Series

Maple can also calculate power series directly for you. Try the commands[2]

$$> \texttt{series}(\exp(x), x = 0);$$
$$> \texttt{series}(\sinh(x), x = 1, 16);$$

(3.53)

The first should return the power-series expansion of the exponential function about $x = 0$ up to but not including 6th order in x. The second command shows how to get more terms: it expands the hyperbolic sine function about $x = 1$ and gives terms up to but not including those of 16th order.

In order to calculate with these series, we must truncate them, taking only terms up to, but not including or higher than, some power. The command `convert(series(), polynom)` approximates the series by a polynomial including all the explicitly calculated terms but omitting the $O(x^n)$ term. Thus

$$> \texttt{y} := \texttt{convert}(\texttt{series}(\sinh(x), x = 1, 16), \texttt{polynom});$$

(3.54)

is the 15th-order polynomial function which approximates the hyperbolic sine function near $x = 1$. The functional expression (3.54) can be converted into a functional operator by "unapplying" it to x:

$$> \texttt{y:=unapply(y,x);}$$

(3.55)

which makes it easy to plot:

$$> \texttt{plot(\{y,sinh\},-3..3);}$$

(3.56)

---

[2]If you receive an error message about the wrong type of parameter, it may be because the variable x was peviously defined. You can free it from its previous association with the commands
> `readlib(unassign);`
> `unassign('x');`
The `readlib` command tells Maple where to find the `unassign` procedure and needs to be called only the first time the procedure is invoked in any given session. Note the quotes needed around the variable to be unassigned.

For comparison purposes, it may be useful to form and plot a sequence of such polynomials. Try, for example, the commands

```
> seq(convert(series(sin(x), x, 2 * n), polynom), n = 1..3);
> plot({", sin (x) }, x = -2..2);
```

$$(3.57)$$

in order to compare three different series approximations for the sine function. Note that if the second argument in the **series** command does not contain an equal sign, the expansion point is given by that argument equal to 0; in other words, the series created is a series in powers of the argument.

> **Exercise 3.10.** Try forming a set of approximations of $\exp(x)$ about $x = 1$ with polynomials of order zero through three in $(x - 1)$. Plot all these approximations together with the exponential function itself over the domain $0 < x < 2$.

3.8   *Limits

Calculations of finite terms in physics often involves indeterminate forms such as 0/0, infinity/infinity, 0*infinity, or infinity-infinity. This is especially true of one of the most successful and best verified theories in physics, namely quantum electrodynamics. The desired finite value can often be found by taking an appropriate limit. Maple is well adapted to finding such limits.

For example, enter the function

$$> \text{f2} := x - > \sin(x)/x + 2 * \cos(x)/x\hat{}2 - 2 * \sin(x)/x\hat{}3; \quad (3.58)$$

and consider its value at $x = 0$. The first term has the form 0/0, and the second and third terms are infinite but of opposite signs. We can find the limiting value as $x \to 0$ by expanding the expression in a power series about $x = 0$, but Maple can also find it directly:

$$> \text{limit}(\text{f2}(x), x = 0); \quad (3.59)$$

Maple should tell you that the result is 1/3. A way to check this result is to note that $f2(x)$ is just the second derivative of

$$-\frac{\sin(x)}{x} = -1 + \frac{x^2}{3!} - \cdots \qquad (3.60)$$

Try the Maple commands

$$\begin{array}{l} > \text{f} := \text{x} - > -\text{sin(x)/x}; \\ > (\text{D@@2})(\text{f})(\text{x}); \end{array} \qquad (3.61)$$

to see that the second derivative of f(x) is really f2(x). The power series of f2(x) can therefore be found by differentiating that for f(x). The leading term in an expansion about $x = 0$ is seen from (3.60) to be $2/3! = 1/3$.

3.9 *Asymptotic series

Power-series expansions of a function $f(x)$ about finite values x_0 of the argument x are often inconvenient if we need to approximate the function at large values of x. An expansion which is not only valid near x_0 but also at $10x_0$ and $1000x_0$ would be preferable. In other words, an expansion "about infinity" would be better than about finite x. To formalize such an expansion, we substitute $x = 1/s$ and perform an expansion in powers of s about $s = 0$. In the final step, we can replace s by $1/x$. The resulting power series in inverse powers of x is called an *asymptotic* expansion.

Consider, for example, the inverse polynomial

$$> \text{p} := 1/(1 - \text{x} - \text{x\^{}2}); \qquad (3.62)$$

The Maple command

$$> \text{pasy} := \text{asympt(p, x)}; \qquad (3.63)$$

generates a power series in inverse powers of x. This is the asymptotic expansion of p. In order to check the accuracy of the asymptotic expansion graphically, we need to eliminate the order-of-magnitude term $O(1/x^6)$ at the end. Unfortunately, the command convert(pasy, polynom) does not seem to work with expansions in inverse powers. However, we can simply subtract the offending term:

$$> \texttt{q := pasy - O(1/x\^{}6) ;} \qquad (3.64)$$

Exercise 3.11. The number of terms in `pasy` is given in Maple by `nops(pasy)`, and the *N*-th term is `op(pasy)[N]`. Write and try a more general Maple command which subtracts the last term from `pasy`.

Answer: `pasy - op(pasy)[nops(pasy)]` .

Exercise 3.12. The Maple function `degree(t)` gives the degree of the term `t`. Check the degree of each term in `pasy` in equation (3.63). Use Maple's on-line help pages to learn about the Maple command **select**, and then explain how the command

$$> \texttt{q := select(t -> degree(t)< 0,pasy);} \quad (3.65)$$

works to produce the result (3.64).

The plot command

$$> \texttt{plot}(\{p, q\}, x = 2..5); \qquad (3.66)$$

then gives the desired comparison.

By default, Maple computes asymptotic expansions up to but not including 6th order in the inverse argument. To perform the calculation of, say, p up to *n*-th order, use the command

$$> \texttt{asympt}(p, x, n); \qquad (3.67)$$

A word of caution is in order about asymptotic expansions: they often do not converge. An approximation based on such an expansion can get better as more terms are added until a minimum error is reached, and thereafter, as still more terms are added, the approximation only gets worse. The magnitudes of the terms in the expansions then follow a similar pattern: successive terms become smaller and smaller until they reach a minimum and then begin increasing. Once they start increasing in size, the expansion is headed for divergence. The larger the value of the argument, the more accurate the expansion and the more terms can be added before it begins to diverge.

3.10 *Problems

1. Find the equation of motion for the pendulum analogous to (3.17) but including terms to and including fourth order in the arc length s. Will the correction to (3.17) tend to increase or decrease the oscillation frequency?

2. Find the smallest power series $y := x \to \sum_n a_n x^n$ which matches the function $\ln(x)$ and its first two derivatives at $x = 1$. To get Maple's help, define a set of matching conditions `cond:={`
`y(1)=ln(1),D(y)(1)=D(ln)(1),(D@@2)(y)(1)=(D@@2)'(ln)(1}`
and use Maple's `solve` command to determine the coefficients a_n. After Maple prints the answer, execute the Maple command
`> assign('');`
to assign those values to the coefficients. With the commands
`> y(s+1);`
`> expand('');`
see how the approximation simplifies when written as a power series in $s = x - 1$. Finally, compare a plot of your polynomial approximation with the ln function itself over the domain 0..5 .

3. Derive the Taylor series for $\sin\theta$ and $\cos\theta$ about $\theta = 0$. To what order must the expansion for $\sin\theta$ be taken in order for $\sin 1$ to be accurate to about 10^{-3}? In order for $\cos 10$ to be that accurate?

4. Find the Taylor series (at least the first six terms) for the function $\exp(-x) * \sin(x)$ and verify that the same series is obtained whether it is calculated directly or determined by multiplying the two series for $\exp(-x)$ and $\sin(x)$ together. Split the function into its even and odd parts, and write these in closed form, that is, NOT as an infinite series.

5. Use the binomial expansion to derive the power-series expansion of $\sqrt{1+x}$ about $x = 0$. Approximate the series on Maple by

summing terms up to 24th and 25th order. By comparing errors estimate the domain of validity (convergence) of the series. State the evidence for the domain you suggest.

6. Ask Maple for the limit of $|x|\cot(x)$ as $x \to 0$. The absolute value $|x|$ of x can be expressed in Maple by abs(x). Does *undefined* mean infinite? Try the commands which take the limit from the right and from the left:
   > limit(abs(x)*cot(x),x=0,right);
   > limit(abs(x)*cot(x),x=0,left);
   To see what's happening, plot the function in the domain $x = -2.5..2.5$.

7. Find the asymptotic expansion of N! for large N. Evaluate the expansion to 3rd order (include the first three terms) for N = 3 and for N = 10. What is the fractional error in each case? (You may need the evalf command.) Repeat the calculation, taking the expansion to (but not including) 6th order. (The first-order asymptotic approximation is known as *Stirling's formula*.)

3.11   Chapter Summary

3.11.1   Concepts:

linear approximations

simple harmonic motion and uniform circular motion

terms of order x^n

power-series approximations

Taylor series expansions

binomial expansions

even and odd functions

sinh and cosh functions

lists and sets

error estimates

products of series

limiting values and asymptotic forms

3.11.2 Maple Commands:

```
seq
```

```
sum
```

```
map
```

```
series
```

```
readlib
```

```
unassign
```

```
sinh, cosh
```

```
convert( ,polynom)
```

```
unapply
```

```
limit
```

```
asympt
```

```
assign
```

```
expand
```

```
abs
```

```
cot
```

```
nops, op
```

```
degree
```

```
select
```

Chapter 4

Vectors

Two- and three-dimensional vectors will be familiar to most readers. Unfortunately, familiarity is not sufficient to prevent applications of vectors in the physical sciences from being a common source of difficulty. Problems often arise as a result of basic misconceptions, fundamental properties of vectors which were not fully appreciated by the student during his/her first introduction to the subject. There is no denying the crucial role played by the vector concept in science; a sound grasp of the fundamentals is mandatory for the successful student. Most students will want to go beyond fundamentals.

One mark of sophistication in the natural scientist is the ability to manipulate vectors as mathematical objects in their own right and not merely as collections of coordinates. This chapter introduces vectors with a notation which is easily extended to geometric algebra and four-dimensional spacetime, as well as to spaces of higher dimensions. Although components on basis vectors are discussed, the emphasis is on vectors as geometric objects. Many vector problems in the physical sciences can be solved without explicit reference to components. Components are convenient when numerical comparisons are made, of course, but they tend to mask the physical relations in a veil of detail. Component-free solutions represent geometrical relations more cleanly.

The approach here is thus to represent the physical quantities such as velocity, acceleration, and force by component-free vectors and to introduce components in conjunction with an observer who carries her/his own reference frame. This approach allows us to consider observers in motion, either at constant velocity or in linear acceleration. Rotating frames of reference will be considered later (in Chapter 9). The applications in this chapter are limited to vectors in flat, Euclidean space, where they are most easily related to familiar concepts, but the presentation is made with an eye toward future applications in Minkowski spacetime and curved manifolds. The worksheet for this chapter contains supplementary material on line plots and on animated and 3-dimensional plots.

Chapter 4 provides essential background for further developments in Chapters 8 and 9, and concepts developed here will also prove useful in the discussions of partial derivatives (Section 5.4) and multivariate data fits (Sections 6.5–6.7).

4.1   *Vectors as Displacements

The concept of points in space is taken as axiomatic. In a three-dimensional Euclidean space, global coordinates can be established on which the positions of points are specified by a collection of three real numbers, and the distance between any two points is well defined. *Vectors* in such a space can be introduced simply as the difference between points. Vectors are usually represented by arrows which join one point to another. Let P_1 and P_2 be two points in the space. The arrow from P_1 to P_2 represents a vector **v** which gives the position of P_2 with respect to P_1 (see Fig. 4.1). We say that **v** gives the *displacement* of P_2 from (or with respect to) P_1. It has both a direction and a magnitude: its orientation gives the direction of P_2 from P_1, whereas its length (magnitude) gives the distance between the two points.

The vector **v** itself preserves no information about the individual points P_1 and P_2; it has only a length (magnitude) and a direction. It gives the relative position not only of P_2 with respect to P_1 but also of any other point P_2' with respect to some P_1' whose relative displacement from P_2' is the same as that of P_1 from P_2. This restriction on the information content of **v** is not meant to deny physical significance to the individual points; the position P_2 by itself may or may not be significant, but the vector **v** says nothing about P_2 except its relative position with respect to P_1. The same vector **v** can thus be represented

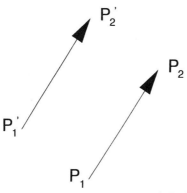

Figure 4.1. The vector **v** is the displacement of P_2 from P_1 as well as of P_2' from P_1'.

by an infinite number of arrows: an arrow can be drawn from any point, but all the arrows are related by *translations* (sliding motions without rotation): all have the same length $v = |\mathbf{v}|$ and the same direction $\hat{\mathbf{v}} = \mathbf{v}/v$. As in Chapter 2, the "hat" indicates a *unit vector*, that is, a vector of length 1; unit vectors are used to indicate directions. Any vector can thus be expressed as a product of a magnitude (its length) and a direction (the unit vector):

$$\mathbf{v} = v\hat{\mathbf{v}}. \tag{4.1}$$

Often we will choose some point O as an origin.[1] Then, any other point P can be associated with a unique *position vector* **r** from O to P. Physical phenomena should not depend on the choice of O; we can make the choice for convenience. For example, we may wish to make O the position of the observer in a measurement.

It is useful to contrast vectors to *scalars*. Scalars signify a positive or negative size, but no direction. A scalar is represented simply by a number. A vector, on the other hand, has both a size (its magnitude is a positive scalar length) and a direction in space. In three-dimensional space, it takes three independent numbers to specify a vector: one

---

[1] A Euclidean space of points is said to be an *affine* space if there is no preferred origin. (However, the concept of an affine space is not restricted to Euclidean vector spaces.)

length and two angles or three Cartesian components (see below), for example. Consequently, a single vector relation gives as much information as three scalar relations.

In Maple, a vector is an array of components; the components are given as a list. Thus, if the vector $\mathbf{v}$ has Cartesian components v_1, v_2, and v_3, it may be specified in Maple by

$$> \ \mathtt{v := array([v1, v2, v3]);} \qquad\qquad (4.2)$$

The components of any vector $\mathbf{v}$ in Maple can be easily extracted; you specify the index of the component in square brackets:

$$> \ \mathtt{v[1], v[2], v[3];}$$

A vector $\mathbf{v}$ can be *scaled* or *dilated* by a positive scalar s to give a new vector $\mathbf{v}' = s\mathbf{v}$ whose direction is the same, but whose length has been multiplied by the factor s. Thus the vector $2\mathbf{v}$ is represented by any arrow which points in the same direction as $\mathbf{v}$ but is twice as long as $\mathbf{v}$. In Maple, we must use the `evalm` command for matrix evaluation of addition (or subtraction) of vectors or multiplication of a vector by a scalar:

$$> \ \mathtt{evalm(2*v);} \qquad\qquad (4.3)$$

Another way to write the vector $2\mathbf{v}$ is as the *vector sum*

$$2\mathbf{v} = \mathbf{v} + \mathbf{v}\,. \qquad\qquad (4.4)$$

In Maple,

$$> \ \mathtt{evalm(v + v);}$$

More generally, the sum of a vector $\mathbf{r}_1$ from the point O to P_1 with another $\mathbf{v}$ from P_1 to P_2 is the vector

$$\mathbf{r}_2 = \mathbf{r}_1 + \mathbf{v} \qquad\qquad (4.5)$$

from O to P_2 (see Fig. 4.2). This example shows how vector addition can be viewed diagrammatically: simply translate the second vector so that it starts at the tip of the first; the sum is the vector from the base of the first to the tip of the second. The same resultant vector can be formed by adding in the reverse order:

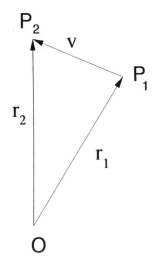

Figure 4.2. The vector *sum* $r_2 = r_1 + v$ is the vector from the base of r_1 to the tip of v where v starts at the tip of r_1. The *vector difference* $v = r_2 - r_1$ is the vector from the tip of r_1 to the tip of r_2 where both r_1 and r_2 start from the same point.

$$r_2 = v + r_1 . \tag{4.6}$$

Therefore vector addition is *commutative*. In a diagram, the two additions (4.5) and (4.6) give a parallelogram (see Fig. 4.3).

We can turn the relation (4.5) around by subtracting r_1 from both sides:

$$v = r_2 - r_1 . \tag{4.7}$$

This new relation is the same one illustrated in Fig. 4.2; it states that the position of P_2 relative to P_1 is a vector difference: the vector r_2 from O to P_2 minus that from O to P_1.

The *zero vector* is the displacement of any point from itself. It has no definable direction and is thus usually represented by the scalar 0. For example, the difference

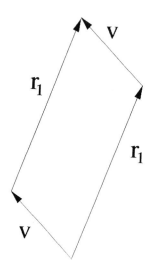

Figure 4.3. In a diagram, the commutivity of vector addition $\mathbf{r}_1 + \mathbf{v} = \mathbf{v} + \mathbf{r}_1$ takes the form of a parallelogram.

$$\mathbf{r}_1 - \mathbf{r}_1 = 0 \tag{4.8}$$

can be considered the *sum* of the vector $\mathbf{r}_1$ from the origin O to P_1 with $-\mathbf{r}_1$ from P_1 back to O:

$$0 = \mathbf{r}_1 + (-\mathbf{r}_1). \tag{4.9}$$

The vector $-\mathbf{r}_1$ is the same as the scalar -1 times $\mathbf{r}_1$. It has the same length as $\mathbf{r}_1$ but the opposite direction. In general, multiplication of a vector by a negative scalar s changes its length by a factor of $|s|$ and reverses its direction.

Once two vectors have been added together, a third vector can be added to the resultant sum. If we think of vectors as displacements from one point to another, it is clear that the result is the same when the first vector is added to the sum of the second and third. If $\mathbf{u}, \mathbf{v}, \mathbf{w}$ are the three vectors, we can write

$$\mathbf{u} + (\mathbf{v} + \mathbf{w}) = (\mathbf{u} + \mathbf{v}) + \mathbf{w} \equiv \mathbf{u} + \mathbf{v} + \mathbf{w}, \tag{4.10}$$

which expresses the *associativity* of vector addition. If the sum of a set of vectors vanishes, this may be interpreted in terms of displacements by saying that after adding all the individual displacements together, you end up where you started. The vectors then form the sides of a closed polygon (see Fig. 4.4).

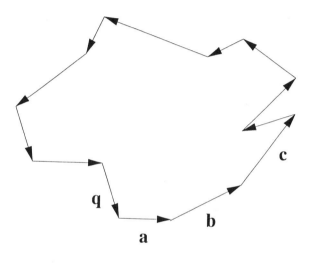

Figure 4.4. A set of vectors whose sum $\mathbf{a} + \mathbf{b} + \mathbf{c} + \ldots + \mathbf{q}$ is zero form a closed polygon.

4.2 Linear Vector Space

Any vector times a scalar is another vector, and the sum of two vectors is also a vector. Any sum of scalar multiples of vectors, called a *linear combination* of vectors, is therefore also a vector. For any real scalars λ, μ, and any vectors $\mathbf{u}$, $\mathbf{v}$ in Euclidean space, linear combinations are seen to obey the following laws:

(i) $\lambda(\mathbf{u} + \mathbf{v}) = \lambda\mathbf{u} + \lambda\mathbf{v}$

(ii) $(\lambda + \mu)\mathbf{u} = \lambda\mathbf{u} + \mu\mathbf{u}$

(iii) $\lambda(\mu\mathbf{u}) = (\lambda\mu)\mathbf{u}$.

Laws (i) and (ii) are known as *distributive* laws, and (iii) is the *associa-*

tive law of scalar multiplication. In particular, the identity for scalar multiplication is 1: $1\mathbf{u} = \mathbf{u}$; $0\mathbf{u} = 0$ gives the zero vector, which is the identity for addition: $0 + \mathbf{v} = \mathbf{v}$. As seen in the last section, vector addition is commutative

(iv) $\mathbf{u} + \mathbf{v} = \mathbf{v} + \mathbf{u}$

and associative

(v) $(\mathbf{u} + \mathbf{v}) + \mathbf{w} = \mathbf{u} + (\mathbf{v} + \mathbf{w})$,

and for every vector $\mathbf{u}$ there is an *additive inverse* $-\mathbf{u}$ which obeys

(vi) $\mathbf{u} + (-\mathbf{u}) = 0$.

Elements adhering to these rules are said to form a *linear vector space.*

The general concept of a vector space is a powerful tool in mathematics; many other entities besides displacements in Euclidean space form vector spaces. Although we concentrate here on vectors in two- and three-dimensional Euclidean space, we discuss a more abstract *parameter space* Chapter 6. In Chapter 9, we introduce an associative vector product, and as we show there, this product is the essential new feature in a *vector algebra*, a rich mathematical structure which contains several vector spaces.

4.3   Motion of Points

If the position $\mathbf{r}$ of a point relative to an origin is given by a constant vector, the point is stationary and its velocity is zero. On the other hand, if the position is a function of time $\mathbf{r} = \mathbf{r}(t)$ given by the product of a constant vector $\mathbf{v}$ and a scalar t representing the elapsed time

$$\mathbf{r}(t) = \mathbf{v}t\,, \tag{4.11}$$

then the direction of the point from the origin is fixed, but its distance increases at a constant rate $v = |\,\mathbf{v}\,|$. The velocity of the point is $\mathbf{v}$ and is not changed if the particle is displaced by a constant vector $\mathbf{r}(0)$:

$$\mathbf{r}(t) = \mathbf{r}(0) + \mathbf{v}t\,. \tag{4.12}$$

This is the vector equation for a straight line parameterized by the scalar quantity t. The *vector function* $\mathbf{r}(t)$ gives the trajectory (or path) of the moving point, passes through the position $\mathbf{r}(0)$ relative to the origin and is aligned along $\mathbf{v}$.

More generally, if the position of a moving point is given by $\mathbf{r}(t)$, its velocity $\mathbf{v}$ is the time-rate of change of the position:

$$\mathbf{v} = \frac{d\mathbf{r}(t)}{dt} \equiv \dot{\mathbf{r}}(t) \equiv \lim_{h \to 0} \left[\frac{\mathbf{r}(t+h) - \mathbf{r}(t)}{h} \right] . \qquad (4.13)$$

It is tangent to the path of the point at $\mathbf{r}(t)$. The time-derivative of the velocity gives the acceleration.

If the vector $\mathbf{r}$ is written as a product $r\hat{\mathbf{r}}$ of length r and direction $\hat{\mathbf{r}}$, its derivative can arise from either a change in magnitude, a change in direction, or both:

$$\dot{\mathbf{r}} = \dot{r}\hat{\mathbf{r}} + r\dot{\hat{\mathbf{r}}} . \qquad (4.14)$$

A point at $\mathbf{r}(t)$ can thus have a nonzero velocity even if its distance from the origin r is fixed. The obvious example is a point in circular motion about the fixed origin O .

As an example, consider motion in a uniform gravitational field. In a uniform gravitational field, every free object has the same constant acceleration $\mathbf{g}$

$$\dot{\mathbf{v}}(t) \equiv \frac{d\mathbf{v}(t)}{dt} = \mathbf{g} . \qquad (4.15)$$

The integral of the acceleration with respect to the time t gives a velocity which is linear in t:

$$\mathbf{v}(t) = \int dt \mathbf{g} = \mathbf{g} \int dt = \mathbf{v}(0) + \mathbf{g} t . \qquad (4.16)$$

The term $\mathbf{v}(0)$ enters as an integration constant but can be seen to be the velocity at $t = 0$. Another integral gives the position

$$\mathbf{r}(t) = \int dt \mathbf{v}(t) = \mathbf{r}(0) + \mathbf{v}(0)t + \frac{1}{2}\mathbf{g} t^2 . \qquad (4.17)$$

Note that this has the form of a power-series expansion of $\mathbf{r}(t)$ about $t = 0$. Here, however, the coefficients are constant *vectors*, and the expression is exact because only the first two derivatives of $\mathbf{r}(t)$ are nonzero.

4.4 *Relative Positions and Velocities

Absolute positions and velocities are rarely if ever important in physics. It is the *relative positions* and relative velocities of objects with respect to each other or with respect to an observer which is physically significant. Often it is possible to fix the origin O with respect to the observer or at one of the objects, but sometimes all the objects and the observer are in motion with respect to the chosen origin.

Consider, for example, an object of mass m at $\mathbf{r}$ and an observer at $\mathbf{R}$. The object responds to a net force $\mathbf{F}$ with an acceleration given by Newton's second law:

$$\ddot{\mathbf{r}} = \mathbf{F}/m. \tag{4.18}$$

The equation is written implicitly with respect to the origin O, but it is also valid in the frame of the observer as long as she moves at constant velocity with respect to the origin: $\ddot{\mathbf{R}} = 0$. Note that $\mathbf{F}$ is the *net* or *total force*. That is, it is the vector sum of all forces on the object.

> **Exercise 4.1.** When a given plane is tilted from the horizontal by an angle θ, a block slides down it at constant velocity. Find the frictional force under these circumstances. What is the coefficient of friction between the block and the plane?
>
> **Solution:** Since the block moves at constant velocity, its acceleration is zero. By (4.18), the net force on it vanishes, and the gravitational force must be exactly balanced by the sum of the normal force $\mathbf{N}$ of magnitude $mg\cos\theta$ perpendicular to the plane and the frictional force $\mathbf{f}$ of magnitude $\mu mg\cos\theta$ up the plane (see Fig. 4.5).
>
> The frictional force is therefore $f = mg\sin\theta = \mu mg\cos\theta$ and the coefficient of friction is simply the slope of the plane: $\mu = \tan\theta$.

Integrals of the *equation of motion* (4.18) give the velocity $\dot{\mathbf{r}}$ and position $\mathbf{r}$ of the object as a function of time. The relative position

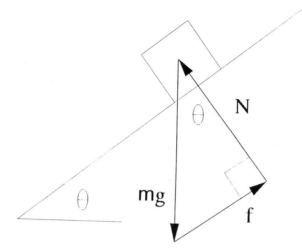

Figure 4.5. The slope $\tan\theta$ at which a block slides down an inclined plane at constant velocity is the coefficient of friction. The sum of the gravitational force $m\mathbf{g}$, the normal force $\mathbf{N}$, and the frictional force $\mathbf{f}$ then vanishes.

with respect to the observer is $\mathbf{r}(t) - \mathbf{R}(t)$. Thus the observer, from her *frame of reference*, sees the object move with velocity $\dot{\mathbf{r}} - \dot{\mathbf{R}}$ and acceleration

$$\ddot{\mathbf{r}}(t) - \ddot{\mathbf{R}}(t) = \mathbf{F}/m - \ddot{\mathbf{R}}(t). \tag{4.19}$$

As long as the observer moves with constant velocity (possibly but not necessarily zero), $\ddot{\mathbf{R}}(t) = 0$, and she sees the object moving in accordance with Newton's second law. If she accelerates, Newton's second law doesn't immediately work. Frames of reference which move at constant velocity (i.e., constant speed and direction)[2] are called *inertial frames*. The familiar laws of physics, such as Newton's second law[3], all hold in inertial frames.

---

[2]Constant with respect to what, you ask? Good question! But once a single inertial frame has been identified, all the rest can identified by reference to it.

[3]For nonrelativistic events, of course.

Exercise 4.2. A monkey sits on a branch waiting for his trainer to fire a banana to him from a small cannon. Unfortunately the trainer is new at this and has aimed the cannon without considering the effect of gravity. Show that if the monkey drops from his branch just as the cannon is fired, he will still be able to catch the banana.

Solution: Both the banana and the monkey fall in the gravitational field with the same acceleration $\mathbf{g}$. Let the position of the banana be $\mathbf{r}(t)$ and that of the monkey $\mathbf{R}(t)$. The trainer has aimed the cannon so that

$$\mathbf{r}(0) + \mathbf{v}(0)t_c = \mathbf{R}(0) \tag{4.20}$$

at some time t_c after the cannon is fired. In the gravitational field, after the monkey drops from his branch, the positions are

$$
\begin{aligned}
\mathbf{r}(t) &= \mathbf{r}(0) + \mathbf{v}(0)t + \frac{1}{2}\mathbf{g}t^2 \\
\mathbf{R}(t) &= \mathbf{R}(0) + \frac{1}{2}\mathbf{g}t^2
\end{aligned}
\tag{4.21}
$$

The difference $\mathbf{r}(t) - \mathbf{R}(t)$ gives the position of the banana with respect to the falling monkey, and since the accelerations of the monkey and banana are the same in a gravitational field, the acceleration of the banana relative to the monkey is zero. The effect of the gravitational field on the relative position thus vanishes; there is no gravitational acceleration in the freely falling frame of the dropping monkey:

$$\mathbf{r}(t) - \mathbf{R}(t) = \mathbf{r}(0) + \mathbf{v}(0)t - \mathbf{R}(0) \tag{4.22}$$

and this goes to zero at t_c.

Exercise 4.3. Show, using vectors, that if two balls are thrown in a uniform gravitational field at the same time from the same position with different initial velocities, that they cannot meet before at least one of them strikes another object.

With Maple's animation feature, trajectories of moving points can be played forward and backward at different speeds. Either you can build up a sequence of plots which you display in rapid succession, or you animate some plots directly. An important example of the latter option is the animation of a traveling wave:

```
> with(plots);
> animate( sin (x − t), x = −3 * Pi..3 * Pi, t = 0..15        (4.23)
> *Pi/8, axes = boxed, scaling = constrained);
```

The same information can also be rendered in a static 3-dimensional plot:

```
> plot3d(sin(x − t), x = −3 * Pi..3 * Pi, t = 0..4 * Pi);     (4.24)
```

The height of the plot gives the functional value, and x and t lie in the horizontal plane. Several other examples of animated plots are given in the worksheet for this chapter.

4.5 Einstein's Principle of Equivalence

As seen above in (4.19), if an observer at position $\mathbf{R}$ is accelerating, she will see the motion of objects obey

$$m\left[\ddot{\mathbf{r}}(t) - \ddot{\mathbf{R}}(t)\right] = \mathbf{F} - m\ddot{\mathbf{R}}(t). \qquad (4.25)$$

She may interpret the observations by saying that Newton's second law holds, but with an additional "force" $-m\ddot{\mathbf{R}}(t)$ in the direction opposite to her acceleration. The added term is not a real force, but rather a *pseudoforce* or a *fictitious force*. If the observer monitors the positions of different objects, she finds that the pseudoforce accelerates them all by the same amount, regardless of their mass. This is an essential characteristic of gravitational forces: all objects irrespective of their masses, experience the same acceleration. Thus, an accelerating reference frame has the same effect as a uniform (but possibly nonconstant) gravitational field. The coincidence was not lost on Albert Einstein, who formulated it as a fundamental principle of general relativity. He called it the *principle of equivalence*, and we can paraphrase it as follows:

> *There is no way to distinguish the effects of a uniform gravitational field from the acceleration of the observer.*

This principle is the key to his derivation of the general theory of relativity, which describes the effects of gravity in terms of space-time curvature.

> **Exercise 4.4.** Suppose you stand on a bathroom scale while riding in an elevator. If your weight, according to the scale, increases by 10%, what is the acceleration of the elevator? What is it if your weight is seen to double? To drop to zero?
>
> **Solution**: The change in apparent weight indicates the presence of a fictitious force. If the scale reading increases by 10%, then the fictitious force must equal one-tenth the usual gravitational force at the surface of the earth. Since that force is directed downward, your acceleration in the elevator (see above) must be upward: $\ddot{\mathbf{R}} = -\mathbf{g}/10$. When the scale shows twice your normal weight, the elevator must be accelerating upward at the acceleration of gravity: $\ddot{\mathbf{R}} = -\mathbf{g}$. When the scale reads 0, the elevator must have an acceleration downward equal to that of gravity: $\ddot{\mathbf{R}} = \mathbf{g}$; in other words, the elevator must be in free fall!
>
> **Exercise 4.5.** Consider a spaceship which simulates the gravitational field on earth by undergoing constant linear acceleration. Find the change in velocity of the spaceship after one year of such acceleration (ignore relativistic effects for this calculation). [Answer: slightly more than the speed of light. Since, in fact, massive objects cannot be accelerated to the speed of light, it is evidently NOT a good approximation to ignore relativistic effects in this problem!]

Exercise 4.6. Describe the trajectory of a jet plane while it is simulating the zero gravity of interstellar space for its passengers.

Answer: Ballistic; that is, the plane will be in free fall during this period. (Why?)

Exercise 4.7. Explain why astronauts in a space shuttle or space station only two or three hundred kilometers above the surface of the earth experience weightlessness. Does the earth's gravitational field vanish there? (No!)

4.6 Linear Independence and Dot Products

If a linear combination of vectors is zero, but the coefficients are not all zero, the vectors are said to be *linearly dependent*. If two vectors $\mathbf{u}$, $\mathbf{v}$ are linearly dependent, they are aligned, that is, they are either parallel or antiparallel. Assume $\mu \neq 0$; the statement of linear dependence $\lambda\mathbf{u} + \mu\mathbf{v} = 0$ implies $\mathbf{v} = -(\lambda/\mu)\mathbf{u}$. The two vectors are parallel if $\lambda/\mu < 0$ and are antiparallel if $\lambda/\mu > 0$.

Vectors which are not linearly dependent are *linearly independent*. Thus if two vectors are linearly independent, they are not aligned, that is, they are not *collinear*.[4] Instead, they define a plane, and any linear combination of them lies in that plane. Conversely, any vector in the plane can be expressed as a linear combination of the two original vectors. Thus there cannot be more than two linearly independent vectors in a plane. In fact, that is what is meant by saying that a plane is two dimensional.

To work out the linear combinations, we need the concept of a *dot product*. The dot product of vectors $\mathbf{a}$, $\mathbf{b}$ is defined by

$$\mathbf{a} \cdot \mathbf{b} = |\mathbf{a}||\mathbf{b}|\cos\theta \qquad (4.26)$$

where θ is the angle between $\mathbf{a}$ and $\mathbf{b}$. Note that the dot product is symmetric in the two vectors; thus it is commutative:

---

[4]Yes, that's the right spelling, as illlogical as it seems!

$$\mathbf{a} \cdot \mathbf{b} = \mathbf{b} \cdot \mathbf{a} \,. \tag{4.27}$$

The dot product of two vectors is a scalar which may be envisioned as the length of $\mathbf{a}$ times the projection of $\mathbf{b}$ in the direction of $\mathbf{a}$, or equivalently as the length of $\mathbf{b}$ times the projection of $\mathbf{a}$ in the direction of $\mathbf{b}$. The dot product is usually defined only between vectors but is itself a scalar; therefore it is not associative.

It is also linear in its two factors. Thus if $\mathbf{b}$ is a linear combination $\mathbf{b} = \lambda \mathbf{u} + \mu \mathbf{v}$ of the vectors $\mathbf{u}$, $\mathbf{v}$, then

$$\mathbf{a} \cdot \mathbf{b} = \mathbf{a} \cdot (\lambda \mathbf{u} + \mu \mathbf{v}) = \lambda \mathbf{a} \cdot \mathbf{u} + \mu \mathbf{a} \cdot \mathbf{v} \,. \tag{4.28}$$

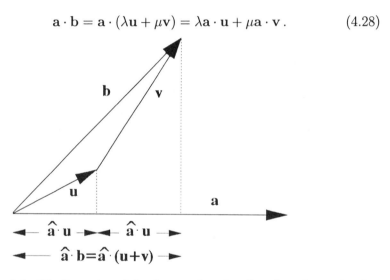

Figure 4.6. The linearity of the dot product implies that the projection of the vector $\mathbf{u} + \mathbf{v}$ in the direction $\hat{\mathbf{a}}$ is the sum of the projections of $\mathbf{u}$ and $\mathbf{v}$ in that direction.

Linearity is shown in Fig. 4.6 for the dot product of a unit vector $\hat{\mathbf{a}}$ with the sum of two vectors $\mathbf{b} = \mathbf{u} + \mathbf{v}$ (this corresponds to the above relation with $\lambda = \mu = 1$). In this case, the linearity of the dot product implies that the projection (or, equivalently, the component) of $\mathbf{b}$ in the direction of $\mathbf{a}$ is the sum of the projections (components) in the same direction of $\mathbf{u}$ and $\mathbf{v}$. The more general relation of linearity follows from $\hat{\mathbf{a}} \cdot (\mathbf{u} + \mathbf{v}) = \hat{\mathbf{a}} \cdot \mathbf{u} + \hat{\mathbf{a}} \cdot \mathbf{v}$ when both sides are multiplied by the magnitude of **a** and the vectors **u**, **v** are replaced by λ**u**, μ**v**, respectively.

If the dot product of two nonzero vectors vanishes, each has zero projection on the other. The angle θ between them is $\pi/2$, and we say they are *perpendicular* or, in other words, *orthogonal* to each other. Thus if $|$ **a** $| \neq 0$ and $|$ **b** $| \neq 0$,

$$\mathbf{a} \cdot \mathbf{b} = 0 \Longleftrightarrow \cos \theta = 0. \tag{4.29}$$

The dot product of a vector with itself is seen to be the square of its length; it vanishes only for the zero vector:

$$| \mathbf{a}^2 | = \mathbf{a} \cdot \mathbf{a}. \tag{4.30}$$

Exercise 4.8. Prove that two vectors are perpendicular iff the length of their vector sum equals the length of their vector difference.

Solution: $| \mathbf{a} \pm \mathbf{b} |^2 = | \mathbf{a} |^2 + | \mathbf{b} |^2 \pm 2\mathbf{a} \cdot \mathbf{b}$, which is short-hand for two equations, one where the upper $(+)$ sign on the left-hand side goes with the upper sign on the right, and one using the lower $(-)$ sign on the two sides. From this, one sees that the square lengths are equal iff $\mathbf{a} \cdot \mathbf{b} = 0$, that is, iff the vectors are perpendicular. Since the lengths of vectors in Euclidean space are real and nonnegative, the square lengths are equal iff the lengths themselves are.

An infinitesimal change in a vector of constant length is always perpendicular to the vector. This can be demonstrated with a pencil on your desk. Put the eraser end on the desk and point the pencil upward. The eraser end is the base of the vector. Now make a small change in the vector. Since you can't change the length of the vector (no chewing or sharpening allowed!), the only change you can make while keeping its base fixed is to swing the point from side to side. The motion of the tip is always perpendicular to the pencil itself.

To see this mathematically, let **a** be the vector and d**a** be a van-ishingly small change which generally points in a different direction than **a**. By using the "differential" symbol d**a** we mean simply that the change is so small that second-order terms in $d\mathbf{a}$ can be ignored (see Chapter 3 on power series expansions). The vector $\mathbf{a} + d\mathbf{a}$ after the change has the same length as before:

$$(\mathbf{a} + d\mathbf{a})^2 = \mathbf{a} \cdot \mathbf{a} + 2\mathbf{a} \cdot d\mathbf{a} = \mathbf{a} \cdot \mathbf{a}. \tag{4.31}$$

We ignore the square length of $d\mathbf{a}$ because it is assumed to be negligible. Consequently, $\mathbf{a} \cdot d\mathbf{a} = 0$ and the change $d\mathbf{a}$ must indeed be perpendicular to $\mathbf{a}$. In particular, the time rate of change of a vector of constant length (for example, a unit vector) must be perpendicular to the vector. The converse is also true: if the time-rate of change of a vector is perpendicular to the vector, there is no change in its length. For example, the force exerted on an electrical charge as it moves with velocity $\mathbf{v}$ through a magnetic field is perpendicular to $\mathbf{v}$. Because the force and hence acceleration of the charge is perpendicular to its velocity, it causes no change in the magnitude of the velocity or the kinetic energy.

This result can be seen in another way. The work done by a force $\mathbf{F}$ acting on an object which is displaced by a vanishingly small amount $d\mathbf{r}$ is given by the dot product

$$dW = \mathbf{F} \cdot d\mathbf{r}. \tag{4.32}$$

The rate of energy transfer from the force to the object is the power

$$P = \frac{dW}{dt} = \mathbf{F} \cdot \mathbf{v}, \tag{4.33}$$

which may be derived by taking a time-derivative of the kinetic energy:

$$\begin{aligned} \frac{d}{dt}\left[\frac{1}{2}m(\dot{\mathbf{r}} \cdot \dot{\mathbf{r}})\right] &= \frac{1}{2}m(\ddot{\mathbf{r}} \cdot \dot{\mathbf{r}} + \dot{\mathbf{r}} \cdot \ddot{\mathbf{r}}) = m\dot{\mathbf{r}} \cdot \ddot{\mathbf{r}} \\ &= \mathbf{v} \cdot m\ddot{\mathbf{r}} = \mathbf{v} \cdot \mathbf{F}. \end{aligned} \tag{4.34}$$

The power transferred vanishes for a charge in a magnetic field. On the other hand, the work done by gravity on an object of mass m which falls a height h is mgh, whereas gravity performs no work on an object which moves horizontally.

Exercise 4.9. Show that the kinetic energy gained by a block of mass m sliding down an inclined plane is $mg(h-\mu s)$ where μ is the coefficient of friction, h is the decrease in height, and s is the horizontal distance moved by the block.

Solution: Recall that the frictional force on the sliding block, directed opposite to the velocity of the block, is μN where N is the magnitude of the normal force of the plane on the block. If the block does not accelerate in the direction normal (perpendicular) to the plane, the projection of the net force in that direction must vanish: $\mathbf{N}+m\mathbf{g}\cdot\mathbf{n} = 0$ where $\mathbf{n}$ is the direction normal to the plane. Consequently, $N = mg\cos\theta$ where θ is the tilt of the plane from the horizontal. The work done by friction as the block slides a distance dr down the plane is thus $dW = -\mu N dr = -\mu mg ds$, where $ds = \cos\theta dr$ is the horizontal component of $d\mathbf{r}$. Integration is trivial, and we find that the energy lost to friction is μmgs (note that it is independent of the height h), whereas the energy gained from the gravitational force is, as above, independent of s: mgh. The net kinetic-energy gain is the difference $mg(h - \mu s)$.

Exercise 4.10. A rabbit hops at constant velocity[5] eastward across a field. It is 30 m directly north of a dog when it is spotted. The dog takes off in hot pursuit, always running with constant speed directly toward the rabbit. How far must the dog run to catch the rabbit if its speed is twice that of the rabbit?

Solution: Let $\mathbf{r}(t)$ be the position of the rabbit and $\mathbf{R}(t)$ that of the dog. The position of the rabbit with respect to the dog is the difference $\mathbf{s}(t) := \mathbf{r}(t) - \mathbf{R}(t)$. Let $\mathbf{v} = v\hat{\mathbf{v}}$ be the constant velocity of the rabbit and V the speed of the dog. The relative velocity is

$$\dot{\mathbf{s}} \equiv \dot{s}\hat{\mathbf{s}} + s\dot{\hat{\mathbf{s}}} = \mathbf{v} - V\hat{\mathbf{s}}. \tag{4.35}$$

---

[5]Averaged over a hop, obviously. (Picky proofreaders!)

It is natural to take components along the perpendicular
directions $\hat{\mathbf{s}}, \dot{\hat{\mathbf{s}}}$:

$$\dot{s} = \mathbf{v} \cdot \hat{\mathbf{s}} - V$$
$$s\dot{\hat{\mathbf{s}}} = \mathbf{v} - \mathbf{v} \cdot \hat{\mathbf{s}}\hat{\mathbf{s}} . \qquad (4.36)$$

Taking another derivative and combining, we obtain

$$s\ddot{s} = s\mathbf{v} \cdot \dot{\hat{\mathbf{s}}} = \mathbf{v} \cdot (\mathbf{v} - \mathbf{v} \cdot \hat{\mathbf{s}}\hat{\mathbf{s}}) = \mathbf{v}^2 - \dot{s}^2 - 2\dot{s}V - \mathbf{V}^2 . \quad (4.37)$$

which is easily rearranged to

$$\frac{d}{dt}(s\dot{s} + 2sV) = v^2 - V^2 . \qquad (4.38)$$

Integration gives an integration constant which is evaluated
by putting $t = 0$ at the start of the chase. At that point
$\dot{s}(0) = -V$ and

$$s(\dot{s} + 2V) = (v^2 - V^2)t + s(0)V \qquad (4.39)$$

The dog catches the rabbit when $s = 0$, and thus the distance covered by the dog is

$$Vt = s(0)\frac{V^2}{V^2 - v^2} . \qquad (4.40)$$

For $v = V/2$, this gives a distance traveled by the dog of
$(4/3)s(0) = 40m$. It's interesting that the distance depends
only on the ratio V/v of two speeds and not on the actual
values.

Dot products are useful for defining planes. Consider the equation

$$\mathbf{r} \cdot \hat{\mathbf{n}} = s \qquad (4.41)$$

where $\hat{\mathbf{n}}$ is a fixed unit vector and s is a positive scalar constant which
gives the projection of $\mathbf{r}$ in the direction $\hat{\mathbf{n}}$. All the positions given by
$\mathbf{r}$ satisfying the above equation with respect to a given origin lie on a
plane perpendicular to $\hat{\mathbf{n}}$ which intersects the direction $\hat{\mathbf{n}}$ at a distance
s from the origin. By saying that the plane is perpendicular to $\hat{\mathbf{n}}$ we
mean that every vector in the plane is perpendicular to $\hat{\mathbf{n}}$. To see

that any vector in the plane defined by the above equation is indeed perpendicular to $\hat{\mathbf{n}}$, consider any two points $\mathbf{r}_1, \mathbf{r}_2$ which satisfy

$$\mathbf{r}_1 \cdot \mathbf{n} = s = \mathbf{r}_2 \cdot \mathbf{n}. \tag{4.42}$$

The displacement of one point from the other defines a vector in the plane which obeys

$$(\mathbf{r}_2 - \mathbf{r}_1) \cdot \mathbf{n} = \mathbf{r}_2 \cdot \mathbf{n} - \mathbf{r}_1 \cdot \mathbf{n} = s - s = 0 \tag{4.43}$$

and is therefore perpendicular to $\hat{\mathbf{n}}$. The vector $\hat{\mathbf{n}}$ is the *unit vector normal to the plane*. It defines the orientation of the plane. If $\hat{\mathbf{n}}$ is vertical, we usually speak of the plane as being horizontal, and conversely if $\hat{\mathbf{n}}$ is horizontal, the plane is usually said to be vertical. Note that if $s = 0$ in (4.41), the origin O itself lies in the plane.

Exercise 4.11. Describe the loci of points $\mathbf{r}$ which satisfy $\mathbf{r} \cdot \mathbf{a} = 3$ where $\mathbf{a}$ is a vector of length 3 pointing along the z-axis.

Now consider the problem of determining what linear combination of linearly independent vectors gives another vector. Given vectors $\mathbf{a}, \mathbf{u}, \mathbf{v}$ and the relation

$$\mathbf{a} = \lambda \mathbf{u} + \mu \mathbf{v}, \tag{4.44}$$

the scalar coefficients λ, μ can be found by taking dot products with $\mathbf{u}$ and $\mathbf{v}$:

$$\begin{aligned} \mathbf{a} \cdot \mathbf{u} &= \lambda u^2 + \mu \mathbf{v} \cdot \mathbf{u} \\ \mathbf{a} \cdot \mathbf{v} &= \lambda \mathbf{u} \cdot \mathbf{v} + \mu v^2. \end{aligned} \tag{4.45}$$

These two simultaneous scalar equations can be solved for the coefficients λ and μ. The method is easily extended to higher-dimensional spaces with more linearly independent vectors.

Exercise 4.12. Let $\mathbf{e}_1, \mathbf{e}_2$ be unit vectors directed along the x- and y-axes, respectively. Express the vector $10\mathbf{e}_2$ as a linear combination of the unit vectors $\mathbf{e}_1$ and $(\mathbf{e}_1 + \mathbf{e}_2)/\sqrt{2}$.

Exercise 4.13. The common molecule methane (CH_4, natural gas, swamp gas) is an example of a tetrahedral molecule. A carbon atom occupies the center, and hydrogen atoms are positioned at the vertices of a regular tetrahedron. The C-H bond directions can be represented by unit vectors $\mathbf{b}_j$, $j = 1, 2, 3, 4$ from the center to each vertex. By symmetry, the angle between any pair of the vectors is the same as between any other pair:

$$\mathbf{b}_1 \cdot \mathbf{b}_1 = \mathbf{b}_2 \cdot \mathbf{b}_2 = \mathbf{b}_3 \cdot \mathbf{b}_3 = \mathbf{b}_4 \cdot \mathbf{b}_4 = 1$$
$$\mathbf{b}_j \cdot \mathbf{b}_k = \beta \equiv \cos\theta \, , \, j \neq k \, . \tag{4.46}$$

Any three of the vectors form a basis and span 3-dimensional space. Therefore, any of the vectors can be expressed as a linear combination of the other three. By symmetry, the expansion coefficients are all the same:

$$\mathbf{b}_1 = \alpha(\mathbf{b}_2 + \mathbf{b}_3 + \mathbf{b}_4)$$
$$\mathbf{b}_2 = \alpha(\mathbf{b}_3 + \mathbf{b}_4 + \mathbf{b}_1)$$
$$\mathbf{b}_3 = \alpha(\mathbf{b}_4 + \mathbf{b}_1 + \mathbf{b}_2) \tag{4.47}$$
$$\mathbf{b}_4 = \alpha(\mathbf{b}_1 + \mathbf{b}_2 + \mathbf{b}_3)$$

Find the expansion coefficient and the angle between the bonds.

Solution: Combine the above equations to give

$$\mathbf{b}_1 \cdot \mathbf{b}_1 = 1 = \alpha(\mathbf{b}_2 \cdot \mathbf{b}_1 + \mathbf{b}_3 \cdot \mathbf{b}_1 + \mathbf{b}_4 \cdot \mathbf{b}_1) = 3\beta\alpha$$
$$\mathbf{b}_1 \cdot \mathbf{b}_2 = \beta = \alpha(\mathbf{b}_2 \cdot \mathbf{b}_2 + \mathbf{b}_3 \cdot \mathbf{b}_2 + \mathbf{b}_4 \cdot \mathbf{b}_2) = \alpha(1 + 2\beta) \, .$$
$$\tag{4.48}$$

Solving these simultaneously for α, β and discarding the solution $\beta = 1$, we find

$$\beta = \cos\theta = -1/3 \, , \, \alpha = -1 \, . \tag{4.49}$$

The interbond angle is thus $\theta = \arccos(-1/3) \approx 109.5° \, .$

4.7 Basis Vectors

Any set of the largest possible number of linearly independent vectors in a given space forms a *basis* for that space. Thus a set of two linearly independent vectors in a plane forms a basis for the plane, and a set of three linearly independent vectors in three-dimensional volume forms a basis there. Since the basis contains the largest number of linearly independent vectors in the space, no vector in the space can be linearly independent of the basis vectors. One says that the basis *spans the space*: every vector must be a linear combination of the basis vectors. The coefficients of the linear combination can be found as in the last section. The simultaneous equations are trivial to solve when an *orthonormal basis* is used. In such a basis, the basis vectors are all unit vectors and are orthogonal to each other.

Let $\{\mathbf{e}_1, \mathbf{e}_2, \mathbf{e}_3\}$ be an orthonormal basis. The orthonormality conditions can be written

$$\begin{aligned} \mathbf{e}_1 \cdot \mathbf{e}_1 = \mathbf{e}_2 \cdot \mathbf{e}_2 = \mathbf{e}_3 \cdot \mathbf{e}_3 = 1 \\ \mathbf{e}_1 \cdot \mathbf{e}_2 = \mathbf{e}_2 \cdot \mathbf{e}_3 = \mathbf{e}_3 \cdot \mathbf{e}_1 = 0 \end{aligned} \tag{4.50}$$

which is more compactly expressed by

$$\mathbf{e}_j \cdot \mathbf{e}_k = \delta_{jk} \tag{4.51}$$

where δ_{jk} is called the Kronecker delta: it is equal to unity if $j = k$ and vanishes otherwise. Any vector $\mathbf{a}$ is a linear combination of the basis vectors:[6]

$$\begin{aligned} \mathbf{a} &= a^1\mathbf{e}_1 + a^2\mathbf{e}_2 + a^3\mathbf{e}_3 \\ &\equiv a^j\mathbf{e}_j . \end{aligned} \tag{4.52}$$

In the second line of the above equation, we introduced the *Einstein summation convention*, whereby repeated indices in an expression are summed over their allowed values. In three-dimensional space,

---

[6]The use of superscripts to represent vector components is particularly useful in relativity and when working with general vector spaces, so it is wise to become acquainted with it now. Its drawback is a possible confusion with the notation for a power, but we will try to ensure that the meaning is clear by context.

$$a^j \mathbf{e}_j \equiv \sum_{j=1}^{3} a^j \mathbf{e}_j \equiv a^1 \mathbf{e}_1 + a^2 \mathbf{e}_2 + a^3 \mathbf{e}_3 \ . \tag{4.53}$$

The coefficients a^j are found by dotting the vector $\mathbf{a}$ into each of the basis vectors:

$$\mathbf{a} \cdot \mathbf{e}_k = a^j \mathbf{e}_j \cdot \mathbf{e}_k = a^j \delta_{jk} = a^k \ . \tag{4.54}$$

Thus

$$a^k = \mathbf{a} \cdot \mathbf{e}_k \ , \quad k = 1, \, 2, \, 3 \ . \tag{4.55}$$

Given a basis, any vector is uniquely determined by its components on the basis vectors. The ordered set of components is called a *representation* of the vector. Thus the vector $\mathbf{a} = a^j \mathbf{e}_j$ can be represented by $(a^1, \, a^2, \, a^3)$. A given vector has many different representations since the representation depends on the basis: a change of basis changes the representations of vectors. The vector itself thus possesses a physical reality and permanence which transcends that of any given representation.

It is often useful to associate the vector itself with the physical quantity of interest and to consider the set of basis vectors as belonging to the *observer*. A given observer can represent the vector by the set of its components which he or she measures on the basis vectors. A different observer will usually have a different representation of the same vector. By taking components, the observer expresses any vector relation as a set of three (assuming three-dimensional vectors) scalar equations, one for each component. If the basis vectors are constant, the change in a vector is due to the change in its components. In particular, if

$$\mathbf{r} = r^k \mathbf{e}_k \tag{4.56}$$

and the $\mathbf{e}_k$ are constant, the time-rate of change of $\mathbf{r}$ is

$$\dot{\mathbf{r}} = \dot{r}^k \mathbf{e}_k \ . \tag{4.57}$$

Vectors in a plane can be expressed as a linear combination of two orthonormal basis vectors, say $\mathbf{e}_1, \mathbf{e}_2$, which span the plane. Let ϕ be the angle between a vector $\mathbf{r} = x\mathbf{e}_1 + y\mathbf{e}_2$ and the basis vector $\mathbf{e}_1$. Then the components of $\mathbf{r}$ can be expressed

$$\begin{aligned} x &= r\cos\phi \\ y &= r\sin\phi \end{aligned} \tag{4.58}$$

where

$$r = \sqrt{x^2 + y^2} \tag{4.59}$$

is the magnitude of $\mathbf{r}$. The pair of variables (r, ϕ) are known as the *polar coordinates* of $\mathbf{r}$.

Consider the example of ballistic trajectories. A common problem treated with components on an orthonormal basis is to find the range of an object on a ballistic trajectory. Suppose the object (ball or missile) was launched with an initial velocity and continues in free fall in the gravitational field. We put the origin at the launching site and ignore air resistance and the rotation of the earth. The position of the object is

$$\mathbf{r}(t) = \mathbf{v}_0 t + \frac{1}{2}\mathbf{g}t^2 \tag{4.60}$$

where $\mathbf{g}$ is the gravitational acceleration and $\mathbf{v}_0$ is the initial $(t = 0)$ velocity. The horizontal (x) and vertical (y) components of this vector equation are

$$x = v_0 t \cos\theta, \ y = v_0 t \sin\theta - \frac{1}{2}gt^2. \tag{4.61}$$

Here θ is the angle made by the initial velocity with respect to the horizontal. The time can be eliminated to give

$$y = x\tan\theta - \frac{g}{2}\left(\frac{x}{v_0\cos\theta}\right)^2, \tag{4.62}$$

and the usual range for flat terrain is obtained when we solve for $y = 0$:

$$x = \frac{2v_0^2}{g}\sin\theta\cos\theta = \frac{v_0^2}{g}\sin 2\theta \ \ at \ y = 0. \tag{4.63}$$

The maximum range is attained by maximizing $\sin 2\theta$, in other words, by making $\theta = \pi/4 = 45°$. The result can be modified to take into account a quadratic variation in the height of the ground. If instead of putting $y = 0$ we use $y = a_1 x + \frac{1}{2}a_2 x^2$, the range is

$$x = \frac{2v_0^2 \cos\theta \, (\sin\theta - a_1 \cos\theta)}{g + a_2 v_0^2 \cos^2\theta} \qquad (4.64)$$

in terms of the slope a_1 at $x = 0$ and the curvature a_2 of the ground.

With the help of the orthonormality condition (4.50), the dot product of two vectors $\mathbf{a}$ and $\mathbf{b}$ is easily expressed in terms of components:

$$\begin{aligned}
\mathbf{a} \cdot \mathbf{b} &= (a^j \mathbf{e}_j) \cdot (b^k \mathbf{e}_k) = a^j b^k (\mathbf{e}_j \cdot \mathbf{e}_k) = a^j b^k \delta_{jk} \\
&= a^j b^j = a^1 b^1 + a^2 b^2 + a^3 b^3 \, .
\end{aligned} \qquad (4.65)$$

Exercise 4.14. Show that a set of six non-crossing surface diagonals of a cube can be chosen to form the edges of a regular tetrahedron, and find the angles between the diagonals.

Solution: The cube has six faces, eight vertices (corners) and 12 edges. The vertices can conveniently be put at the positions $(0,0,0)$, $(0,1,0)$, $(0,1,1)$, $(0,0,1)$, $(1,0,0)$, $(1,1,0)$, $(1,1,1)$, and $(1,0,1)$. Consider the six face diagonals connecting any two of the four vertices $(0,0,0)$, $(0,1,1)$, $(1,0,1)$, $(1,1,0)$. These diagonals can be associated with the six vectors

$$\begin{aligned}
(0,1,1) - (0,0,0) &= (0,1,1), & (0,1,1) - (1,1,0) &= (-1,0,1), \\
(1,1,0) - (0,0,0) &= (1,1,0), & (0,1,1) - (1,0,1) &= (-1,1,0), \\
(1,0,1) - (0,0,0) &= (1,0,1), & (1,0,1) - (1,1,0) &= (0,-1,1).
\end{aligned}$$

Since all have the same length, namely $\sqrt{2}$, and have the same magnitudes of dot product with respect to one another, namely 1 or 0, the tetrahedron must be regular. The angle between edges of the tetrahedron which are connected at a vertex is $\arccos(\pm 0.5) = 60°$ or its supplement, $120°$. Thus each face of the tetrahedron is an equilateral triangle. (See also problem 2 and the worksheet for this chapter, where the cube and the inscribed tetrahedron are plotted.)

4.8 Cross Products

Another type of product between vectors is useful for work with vectors in 3-dimensional space. The cross product[7]

$$\mathbf{c} = \mathbf{a} \times \mathbf{b} \qquad (4.66)$$

is defined as a vector perpendicular to the plane of $\mathbf{a}$ and $\mathbf{b}$ whose magnitude is

$$|\mathbf{c}| = |\mathbf{a}|\,|\mathbf{b}|\sin\theta \qquad (4.67)$$

where θ is the angle between $\mathbf{a}$ and $\mathbf{b}$. Of course, if $\mathbf{c}$ is normal to the plane, so is $-\mathbf{c}$. The direction of $\mathbf{c}$ is defined as the direction a right-handed screw would move when turned from $\mathbf{a}$ to $\mathbf{b}$.

By this definition, the cross product is seen to be *noncommutative*. In fact,

$$\mathbf{a} \times \mathbf{b} = -\mathbf{b} \times \mathbf{a}. \qquad (4.68)$$

Not only is the cross product not commutative, it changes sign when the order of factors is reversed. We say that the cross product is *anti-commutative* and note that the cross product of any vector with itself is automatically zero. The cross product of two vectors $\mathbf{u}$ and $\mathbf{v}$ vanishes iff the two vectors are aligned (parallel or antiparallel) and hence proportional: $\mathbf{u} = \lambda\mathbf{v}$, where λ is a scalar. The anticommutivity of the cross product contrasts to the commutivity of the dot product. The dot product of a vector with itself is zero iff the vector itself vanishes. Recall, however, that the dot product of two vectors is not associative because it is a scalar, not a vector. The cross product of vectors is a vector, but it is still not associative. For example, $\mathbf{a} \times (\mathbf{a} \times \mathbf{b})$ is a vector of length $|\,\mathbf{a}\,|^2|\,\mathbf{b}\,|\sin\theta$ in the plane of $\mathbf{a}$ and $\mathbf{b}$, whereas $(\mathbf{a} \times \mathbf{a}) \times \mathbf{b}$ is the zero vector. The cross product can be shown to be linear:

$$\mathbf{a} \times (\lambda\mathbf{u} + \mu\mathbf{v}) = \lambda\mathbf{a} \times \mathbf{u} + \mu\mathbf{a} \times \mathbf{v}.$$

---

[7]Why is it called a cross product? You'd be cross, too, if you were nonassociative, anticommutative, and noninvertible! See below.

It is worth emphasizing that cross products are vectors only in 3-dimensional space. In two dimensions, there is no direction normal to a plane[8], and in four (or more) dimensions there is a whole plane (or hypersurface) of vectors normal to a given plane.

The magnitude of $\mathbf{a} \times \mathbf{b}$ equals the area of the parallelogram formed by $\mathbf{a}$ and $\mathbf{b}$. It is twice the area of the triangle formed by the sides $\mathbf{a}$, $\mathbf{b}$, and $\mathbf{b} - \mathbf{a}$.

The cross product of two vectors is a vector whose components in a given basis can be related to the components of the vector factors if the cross products of the basis vectors themselves are known. An orthonormal basis in 3-dimensional space can be either right-handed or left-handed. If it is *right-handed*,

$$\begin{aligned} \mathbf{e}_1 \times \mathbf{e}_2 &= \mathbf{e}_3 \\ \mathbf{e}_2 \times \mathbf{e}_3 &= \mathbf{e}_1 \\ \mathbf{e}_3 \times \mathbf{e}_1 &= \mathbf{e}_2 \,. \end{aligned} \tag{4.69}$$

If, on the other hand, there are minus signs on one side of all these relations, the basis (or coordinate system) is said to be *left-handed*. It is conventional to use right-handed bases. Expanding vectors in the right-handed basis $\{\mathbf{e}_j\}$ and using the summation convention, we can write the cross product as

$$\mathbf{c} = c^i \mathbf{e}_i = \mathbf{a} \times \mathbf{b} = a^j b^k \mathbf{e}_j \times \mathbf{e}_k \,. \tag{4.70}$$

The components c^i are thus

$$\begin{aligned} c^i &= a^j b^k \mathbf{e}_i \cdot (\mathbf{e}_j \times \mathbf{e}_k) \\ &\equiv a^j b^k \epsilon_{ijk} \end{aligned} \tag{4.71}$$

where $\epsilon_{ijk} := \mathbf{e}_i \cdot (\mathbf{e}_j \times \mathbf{e}_k)$ is called the *Levi-Civita* symbol. Note that ϵ_{ijk} vanishes if any two of the indices are equal, and

$$\begin{aligned} \epsilon_{123} &= \epsilon_{231} = \epsilon_{312} = 1 \\ \epsilon_{321} &= \epsilon_{213} = \epsilon_{132} = -1 \,. \end{aligned} \tag{4.72}$$

---

[8]Sometimes in 2-dimensional space, a scalar quantity is defined whose magnitude is the length of the cross product between the corresponding two vectors in three dimensions. (See also Chapter 8, Problem 2.)

Combining the above relations, we obtain the cross product itself in terms of components:

$$\mathbf{a} \times \mathbf{b} = \mathbf{c} = c^j \mathbf{e}_j = a^j b^k \epsilon_{ijk} \mathbf{e}_i . \qquad (4.73)$$

Exercise 4.15. Write out the terms implicit in the summation convention for the expression $a^j a^k \epsilon_{ijk}$ where a^j are vector components, and prove that the expression is identically zero.

Exercise 4.16. Show that the magnitude of the scalar quantity $\mathbf{a} \cdot (\mathbf{b} \times \mathbf{c})$ is the volume of the parallelepiped formed by the vectors $\mathbf{a}$, $\mathbf{b}$, $\mathbf{c}$.

Cross products are used to define the *angular momentum* $\mathbf{L} = \mathbf{r} \times \mathbf{p}$ of a particle orbiting about the origin where $\mathbf{p} = m\dot{\mathbf{r}}$ is the usual momentum. Since $\dot{\mathbf{r}} \times \dot{\mathbf{r}} = 0$, the time-rate of change of $\mathbf{L}$ is

$$\dot{\mathbf{L}} = \mathbf{r} \times \dot{\mathbf{p}} = \mathbf{r} \times \mathbf{F} \equiv \mathbf{N} , \qquad (4.74)$$

where we used Newton's second law ($\dot{\mathbf{p}} = m\ddot{\mathbf{r}} = \mathbf{F}$) for particles of fixed mass m. The vector $\mathbf{N} = \mathbf{r} \times \mathbf{F}$ is called the *torque* and gives the twisting action of the applied force. When $\mathbf{F}$ is aligned with $\mathbf{r}$, the torque vanishes and the angular momentum $\mathbf{L}$ is *constant* (that is, its time-rate of change is zero). When the direction of $\mathbf{L}$ is fixed, so is the orbital plane containing $\mathbf{r}$ and $\mathbf{p}$. The magnitude of $\mathbf{L}$ is $2m$ times the rate $\frac{1}{2} | \mathbf{r} \times \dot{\mathbf{r}} |$ at which the orbital area is swept out (see Fig. 4.7). It is fixed for planets moving about a massive sun (this is Kepler's second law), for example, because the gravitational force $\mathbf{F}$ is aligned with the position vector $\mathbf{r}$ and the torque therefore vanishes.

Exercise 4.17. Show that the angular momentum of a particle orbiting in the $\mathbf{e}_1 \mathbf{e}_2$ plane is given in polar coordinates by

$$\mathbf{L} = mr^2 \dot{\phi} \mathbf{e}_3$$

and that the kinetic energy $\frac{1}{2} m v^2$ is given by $\frac{1}{2} m \dot{r}^2 + \mathbf{L}^2 / (2mr^2)$.

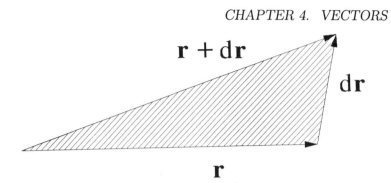

Figure 4.7. The area swept out in time dt is $\frac{1}{2} \mid \mathbf{r} \times d\mathbf{r} \mid$ where $d\mathbf{r} = \dot{\mathbf{r}}dt$. In the figure, this area is the hatched triangle. Of course $d\mathbf{r}$ is greatly exaggerated in the drawing.

4.9   *With(linalg)

Maple has a powerful package for handling vectors and matrices. To invoke the package, issue the command

$$> \texttt{with(linalg);} \qquad\qquad (4.75)$$

The procedures in this package work with vector spaces of arbitrary size. Another package for dealing specifically with vectors of physical space and spacetime is discussed later in Chapter 9.

Vectors and matrices (or rather their representations) can be defined in Maple with the `array` command.[9] Suppose, for example we are given the three linearly independent vectors

$$\begin{aligned}
\mathbf{a}_1 &:= (0,1,1)\mathbf{e} \\
\mathbf{a}_2 &:= (1,0,1)\mathbf{e} \\
\mathbf{a}_3 &:= (1,1,0)\mathbf{e}
\end{aligned} \qquad\qquad (4.76)$$

where $\mathbf{e}$ is the column of basis vectors

$$\mathbf{e} = \begin{pmatrix} \mathbf{e}_1 \\ \mathbf{e}_2 \\ \mathbf{e}_3 \end{pmatrix}. \qquad\qquad (4.77)$$

---

[9]In the `linalg` package, the term `vector` can also be used for a one-dimensional array, and the term `matrix` means a two-dimensional array.

Defining the column

$$
\mathbf{a} := \begin{pmatrix} a_1 \\ a_2 \\ a_3 \end{pmatrix}, \tag{4.78}
$$

we can write the matrix equation

$$
\mathbf{a} = \mathbf{M}\mathbf{e} \tag{4.79}
$$

where the matrix $\mathbf{M}$ is

$$
\mathbf{M} := \begin{pmatrix} 0 & 1 & 1 \\ 1 & 0 & 1 \\ 1 & 1 & 0 \end{pmatrix}. \tag{4.80}
$$

The basis vectors e_1, e_2, e_3 can be expressed in terms of the a_k vectors if we invert the matrix $\mathbf{M}$. We can let Maple perform the inversion for us:

$$
\begin{array}{l} > \texttt{M := array([[0, 1, 1], [1, 0, 1], [1, 1, 0]]);} \\ > \texttt{Minv := inverse(M);} \end{array} \tag{4.81}
$$

The rows of Minv are the vectors e_1, e_2, e_3 in the basis $\{a_1, a_2, a_3\}$.[10] The linalg package also contains commands for finding the dot product of vectors (dotprod or innerprod), for calculating the cross product (crossprod) of vectors in 3-dimensional space, for taking traces (trace) and determinants (det), for finding eigenvalues (eigenvals), and for solving matrix equation (linsolve). Look up the linalg package on Maple's on-line help for more details about these (and many more) commands.

---

[10] Another way to determine the inverse of matrix, one that does not require the linalg package explictly, is to use

$$
> \texttt{evalm}(1/M);
$$

4.10 Problems

1. A rabbit hops across a field, and as it passes a distance $\mathbf{r}(0)$ from a dog, the dog takes chase. Suppose that the dog always runs directly toward the rabbit. The rabbit, on the other hand, instead of running in a straight line, always runs in a direction perpendicular to its line of sight to the dog. Prove that the time it takes the dog to catch the rabbit is independent of how fast the rabbit hops. [Note that the motion of the rabbit does not contribute to changes in the distance between the animals.]

2. Find the angles between body diagonals of a cube. (The four body diagonals join opposite corners and pass through the center of the cube.) Show why one of these angles is also the angle between any two lines joining the center of a regular tetrahedron with its vertices.

3. Find the angle which gives the largest range when firing a ballistic missile up a slope $\Delta y / \Delta x = a_1$. Find the optimal numerical values of θ for the extreme slopes of $45°$ ($a_1 = 1$) and $-45°$ ($a_1 = -1$). [One of the answers is $\pi/8$.]

4. Consider three vectors which sum to zero: $\mathbf{a} + \mathbf{b} + \mathbf{c} = 0$.

   (a) Write down an example of three such vectors in the xy plane and sketch them in a diagram.

   (b) Express the area of the related triangle as three different cross products of vectors.

   (c) Use the equality of the above products to prove the relation

   $$\frac{\sin \alpha}{a} = \frac{\sin \beta}{b} = \frac{\sin \gamma}{c}. \qquad (4.82)$$

   Be sure to define all quantities.

5. Show that $a \cdot (b \times c) = a^i b^j c^k \epsilon_{ijk}$ and that the same result is given

by the *determinant*

$$\begin{vmatrix} a^1 & a^2 & a^3 \\ b^1 & b^2 & b^3 \\ c^1 & c^2 & c^3 \end{vmatrix} = a^1 \left(b^2 c^3 - c^2 b^3 \right) + a^2 \left(b^3 c^1 - c^3 b^1 \right) + a^3 \left(b^1 c^2 - c^1 b^2 \right) .$$

$$(4.83)$$

Use one of these expressions to prove that

$$\mathbf{a} \cdot (\mathbf{b} \times \mathbf{c}) = \mathbf{b} \cdot (\mathbf{c} \times \mathbf{a}) = \mathbf{c} \cdot (\mathbf{a} \times \mathbf{b}). \qquad (4.84)$$

6. Let $\mathbf{a}$, $\mathbf{b}$ be fixed vectors obeying $\mathbf{a} \cdot \mathbf{b} = 0$. Show that the equation $\mathbf{r} \times \mathbf{b} = \mathbf{a}$ determines positions $\mathbf{r}$ which lie on a straight line. Describe the orientation and position of the line. In particular, how close does it come to the origin?

7. Consider the double cross product $\mathbf{s} = \mathbf{a} \times (\mathbf{b} \times \mathbf{c})$ of three vectors, $\mathbf{a}, \mathbf{b}$, and $\mathbf{c}$.

   (a) Show that the result must lie in the plane of $\mathbf{b}$ and $\mathbf{c}$. Write it in the form $\mathbf{s} = \beta \mathbf{b} + \gamma \mathbf{c}$.

   (b) Use the relation $\mathbf{s} \cdot \mathbf{a} = 0$ to determine the ratio $\beta \backslash \gamma$.

   (c) Show that the values of the scalars β and γ are independent of the component of $\mathbf{a}$ perpendicular to $\mathbf{b}$ and $\mathbf{c}$. In other words, they must be determined by the inner products $\mathbf{a} \cdot \mathbf{b}$ and $\mathbf{a} \cdot \mathbf{c}$.

   (d) The double cross product is said to be *multilinear*, that is, linear in each one of its factors. Thus, if $\mathbf{a}$ is multiplied by the scalar λ, this has the effect of multiplying the product $\mathbf{s}$ by the same factor. Apply this fact with part (c) to argue that β must be proportional to $\mathbf{a} \cdot \mathbf{c}$, and γ, to $\mathbf{a} \cdot \mathbf{b}$.

   (e) Combine the results of parts (b) and (d) with a special case to show the general result:

   $$\mathbf{a} \times (\mathbf{b} \times \mathbf{c}) = (\mathbf{a} \cdot \mathbf{c}) \mathbf{b} - (\mathbf{a} \cdot \mathbf{b}) \mathbf{c} . \qquad (4.85)$$

   (This is not the most direct derivation of the result, but it may help you remember the relation.)

8. Use the relation (4.85) to show that given a direction (unit vector) $\hat{\mathbf{n}}$, any vector $\mathbf{v}$ can be split

$$\mathbf{v} = \mathbf{v}_\parallel + \mathbf{v}_\perp \qquad (4.86)$$

into parts $\mathbf{v}_\parallel = (\mathbf{v} \cdot \hat{\mathbf{n}})\,\hat{\mathbf{n}}$ parallel and $\mathbf{v}_\perp = \hat{\mathbf{n}} \times (\mathbf{v} \times \hat{\mathbf{n}})$ perpendicular to $\hat{\mathbf{n}}$.

9. Let $\{\mathbf{e}_1, \mathbf{e}_2, \mathbf{e}_3\}$ be a non-orthonormal basis. Any vector $\mathbf{v}$ can be expanded in the basis

$$\mathbf{v} = v^j \mathbf{e}_j\,, \qquad (4.87)$$

but the coefficients are no longer the components of $\mathbf{v}$ on $\mathbf{e}_j$.

(a) With the help of equation (4.84) show that, instead,

$$v^1 = \mathbf{v} \cdot \mathbf{e}^1 := \frac{\mathbf{v} \cdot (\mathbf{e}_2 \times \mathbf{e}_3)}{\mathbf{e}_1 \cdot (\mathbf{e}_2 \times \mathbf{e}_3)} \qquad (4.88)$$

with other coefficients given by cyclic permutations of this relation. The basis $\{\mathbf{e}^1, \mathbf{e}^2, \mathbf{e}^3\}$ is said to be *reciprocal* to $\{\mathbf{e}_1, \mathbf{e}_2, \mathbf{e}_3\}$.

(b) Demonstrate that the basis vectors obey

$$\mathbf{e}^j \cdot \mathbf{e}_k = \delta^k_j\,. \qquad (4.89)$$

4.11   Chapter Summary

4.11.1   Concepts

vectors

vector addition

scalar multiplication

linear vector spaces

relative positions, relative velocities

derivatives of vectors

equations of motion

frame of reference

inertial frames

Einstein's Principle of Equivalence

linear independence

dot products

basis vectors

right- and left-handed orthonormal basis systems

vector components, projections

cross products

commutivity

associativity

invertibility

Kronecker delta

Levi-Civita symbol, determinant

trajectories

work

angular momentum

torque

Kepler's second law

representations

4.11.2   Maple Commands

```
with(linalg)

array, vector, matrix

inverse

dotprod, innerprod, crossprod

det

trace

eigenvals

linsolve

plot3d

animate
```

Chapter 5

Basic Data Analysis and Statistics

Progress in the physical sciences is based on an interplay between theory and experiment. Mathematical models designed by theorists must predict or explain measurements made by experimentalists. Revolutions in science have often begun with small but stubborn discrepancies between theory and measurement. On the other hand, the most embarrassing moments in the history of science have often arisen as a bandwagon of research grew around a reported observation of new and unexpected effects which, upon closer scrutiny, were revealed as a misinterpretation of data. It is obviously important to reap the maximum information from experimental data while understanding the limitations of the harvest.

This chapter discusses how measurements are analyzed to give best estimates of the measured quantities and how the precision of the resulting data can be expressed and determined. It starts with a review of the difference between accuracy and precision and the relation of these to systematic and random errors. It also reviews the propagation of errors in expressions with several error sources. The concentration in this chapter is on the measurement of a single variate; multivariate analysis and curve fitting will be the topics of Chapter 6. The concept of a distribution is introduced, and the distinction is emphasized between the properties of a measured sample and those of the underlying population; we show how to estimate the mean and variance of the population from the measured values of the sample. In addition to the normal (Gaussian) distribution, we discuss two common discrete distributions, namely the binomial and Poisson distributions.

Maple serves us admirably not only by speeding up the computations and plotting, but also by allowing us to select samples from underlying distributions and compare their properties. We make use of the `stats` package, which has undergone a major revision in Maple V Release 3. Because the changes are substantial, two versions of the worksheet have been made available for this chapter, `ws5r2.ms` for release 2 and `ws5r3.ms` for release 3. We concentrate here on concepts and on commands common to the two releases.

Books have been written on data analysis, probability, and statistics.[1] The coverage in this chapter is obviously quite limited and cannot replace a more thorough study of the subject. Our aim is to reinforce the data-treatment methods taught in introductory physical-science laboratories; to extend it to other experimental situations commonly faced in other undergraduate laboratory work; to provide some theoretical background to the methods, particularly in the understanding of distributions; and to use Maple in the illustration of concepts and in the computation and display of results.

5.1 Accuracy and Precision

An essential property of scientific research is the repeatability of observation. If the same experiment is performed on similar systems, similarly prepared, the results should be the same. Well, nearly the same: it's those aspects of the measurements which agree or at least are consistent with each other which are usually taken to be scientifically significant. In some cases, such as the study of chaos, the details observed may be totally different from one measurement to the next, but there will still be common aspects, such as the onset of period doubling or the positions and characteristics of stability points, which are both consistent and significant. We'll look at this more in Chapter 8.

---

[1]See, for example, John R. Taylor, *An Introduction to Error Analysis* (Oxford University Press, 1982) and R. L. Plackett, *An Introduction to the Theory of Statistics* (Barnes & Noble, Inc. 1971).

When a physical quantity is being measured, the same experiment may produce slightly different numbers when applied to different but similar systems. Only the part that agrees can be significant. More generally, when an experiment is repeated many times on similar systems, a *distribution* of values will usually be found. If all the experiments are, *a priori*, equally valid, the best estimate of the result is the average value, and the *precision* of the result can be estimated from the width of the distribution and the number of measurements made. In a high precision experiment, the measurements will agree to many figures and the distribution of results will be very narrow. A very precise experiment can still be inaccurate, however. See Fig. 5.1 for an example of the results of precise but inaccurate target shooting.

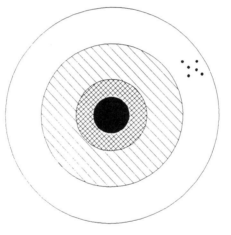

Figure 5.1. The pattern of holes in the target of a firing range shows closely placed shots some distance from the bullseye. It indicates precise but inaccurate marksmanship.

There are many ways that errors can enter an experiment, but we divide all errors into two classes: (1) *random errors*, which are statistical fluctuations in the measurement process or from a distribution of systems arising, perhaps, from an uncontrolled parameter; and (2) *systematic errors*, which can arise from a miscalibration of the measuring instrument or from an ignored influence. The random errors tend to have nearly symmetrical distributions: roughly as many random errors increase the measured value as decrease it. The systematic errors, on the other hand, are more likely to bias the result in one direction or the other. The precision of a result depends on the random errors; it can be quite high if the random errors are well controlled. The *accuracy* of the result, however, depends also on the systematic errors.

If you measure a meter-long rod with a tape measure that has been stretched by 1%, you may be able to achieve a precision of 0.5 mm, that is, of 0.05%, but your result will still be inaccurate. If you ignored the possible systematic error due to the poorly calibrated tape measure, you might give the length of the rod as 990.0 ± 0.5 mm: precise but not accurate. Another common way to express the precision is through the number of significant digits given; the number 0.990 m implies a random error of no more than ± 0.0005 m. Still another method sometimes used is to put the uncertainty in brackets following the digits affected; the above precision could be indicated by 990.0(5) mm.

Obviously, error estimates can be misleading if they do not include systematic errors. First of all, the experimenter needs to be aware of all possible factors which could influence her/his measurement. Then she/he must estimate the size of the influences, reduce the sources of large error, and finally include an estimate of the systematic errors in the final results. Because systematic errors are often biased, their inclusion may lead to asymmetrical error estimates, such as $999.0^{+2.5}_{-0.5}$ mm. The control of systematic errors requires imagination and diligence.

The treatment of random error is somewhat more straightforward. It forms the main topic of the remainder of this chapter.

5.2   Averages and rms Deviations

Consider a simple experiment, such as measuring the length of a rod, in which only one quantity is measured. Because of the random error, different trials (measurements) give slightly different results. Let X represent the length of the rod; in statistical jargon, it is known as the *variate* or *random variable*. Let the value of X measured in the m-th trial be $x^{(m)}$, and let n such trials constitute the experimental *sample*.

The *average* or *mean* value of X in the experiment is defined by

$$\langle X \rangle := \frac{1}{n} \sum_{m=1}^{n} x^{(m)}. \tag{5.1}$$

The notation $\overline{X} = \langle X \rangle$ is used by many authors. Note that "taking the average value" is a linear operation: if a and b are the same in all measurements,

$$\langle aX + b \rangle = a \langle X \rangle + b. \tag{5.2}$$

We will assume that systematic errors have been eliminated and, that as n becomes large, the average value $\langle X \rangle$ approaches the exact value. The deviation of a measured value $x^{(m)}$ from the mean in any trial is simply the difference $x^{(m)} - \langle X \rangle$. The average deviation from the mean is identically zero:

$$\langle X - \langle X \rangle \rangle = \langle X \rangle - \langle X \rangle = 0. \tag{5.3}$$

A measure of the spread of values about the mean is given by the average square deviation (from the mean):

$$\begin{aligned} \langle (X - \langle X \rangle)^2 \rangle &= \langle X^2 \rangle + \langle X \rangle^2 - 2 \langle X \rangle \langle X \rangle \\ &= \langle X^2 \rangle - \langle X \rangle^2 . \end{aligned} \tag{5.4}$$

It is known as the *mean square deviation* from the mean and its square root is the *rms (root mean square) deviation*. The rms deviation is also known as the *standard deviation in the sample* and its square is the *sample variance*.

It is the limit of these quantities after a large number of trials that is physically significant. In this limit, the standard deviation in the sample becomes simply the *standard deviation* σ_X of X:

$$\sigma_X = \lim_{n \to \infty} \langle (X - \langle X \rangle)^2 \rangle^{1/2} , \tag{5.5}$$

and its square becomes the *variance* σ_X^2 of X.[2] Note that the average of the squared value $\langle X^2 \rangle$ is generally greater than or equal to the square of the average value $\langle X \rangle^2$. They are equal if and only if every measurement of X yields the same value, namely the average $\langle X \rangle$.

---

[2]To distinguish these quantities from those for a given experimental sample, σ_X and σ_X^2 are sometimes called the standard deviation and the variance *of the population*. See Section 5.7 for more details.

Exercise 5.1. Consider three measurements of X with values $x^{(1)} = 0$, $x^{(2)} = 2$, $x^{(3)} = 10$. Calculating the required values in your head, compare the average value $\langle X \rangle$, the average of the square $\langle X^2 \rangle$, the square of the average $\langle X \rangle^2$, and the mean square deviation.

Answer: 4, $34\frac{2}{3}$, 16, and $18\frac{2}{3}$.

The standard deviation σ_X is a measure of the likely random error in any single measurement of X. As we will see below, a measurement of X has roughly a 68% probability of being within σ_X of the correct value, a 95% probability of being within $2\sigma_X$, and a 99.7% probability of lying within $3\sigma_X$.

5.3 *Maple's 'stats' Package

Much of Maple's power lies in over 20 packages, each one a collection of procedures related to a single topic. In the last chapter, we used the `plots` package. In this chapter, we use several of the commands in the `stats` package. Although a single command from a package can be invoked by a call of the form

$$> \text{package[command]}; \tag{5.6}$$

it is simpler to use the command

$$> \text{with(package)}; \tag{5.7}$$

which allows one to access any command in the package directly. Thus, to use commands in the `stats` package, type

$$> \text{with(stats)}; \tag{5.8}$$

This is sufficient in Release 2, but in Release 3, the `stats` package contains subpackages which must also be loaded, and you will need to add the command

$$> \text{with(describe)}; \tag{5.9}$$

for the procedures we want to access. Try

$$\text{mean}([2, 0, 1, 2, 0]); \tag{5.10}$$

which should yield the average value, namely 1, of the sample of numbers given in the list. Of course a sequence or a data list can be defined independently, such as

$$\begin{aligned}
&> \ \texttt{s} \ := \ 4, 5, 6, 5; \\
&> \ \texttt{L} := \ [\texttt{s}] \\
&> \ \texttt{av\_s} \ := \text{mean}([\texttt{s}]); \\
&> \ \texttt{av\_s} := \text{mean}(\texttt{L});
\end{aligned} \tag{5.11}$$

The list L=[s] is the *sample* and av_s is the *sample mean*. To compute the mean square deviation of the sample from its mean, we can use the map command introduced in Section 3.6:

$$\begin{aligned}
&> \texttt{sqdev} := \text{map}(\texttt{x}- > (\texttt{x} - \texttt{av\_s})\texttt{\^{}}2, [\texttt{s}]); \\
&> \texttt{msd\_s} := \text{mean}(\texttt{sqdev});
\end{aligned} \tag{5.12}$$

Recall that the map command applies the operation specified in its first argument to every element of the list in the second argument. It can be used to find the mean of each data list in a list of data lists. Thus, if the data sample contains measurements of three properties X, Y, Z of four objects with values

$$\begin{aligned}
&> \texttt{xdat} := [0, 4, 8, 12]; \\
&> \texttt{ydat} := [1, 5, 9, 13]; \\
&> \texttt{xdat} := [2, 6, 10, 14];
\end{aligned} \tag{5.13}$$

then the command

$$> \text{map}(\text{mean}, [\texttt{xdat}, \texttt{ydat}, \texttt{zdat}]); \tag{5.14}$$

should yield the list of average values

$$[6, 7, 8]. \tag{5.15}$$

Further Maple commands specific to Release 2 and Release 3 are illustrated in the two worksheets for this chapter, ws5r2.ms for Release 2 and ws5r3.ms for Release 3.

5.4 *Partial Derivatives

To derive the results of the next section, we need to extend the concept of differentiation to functions of several variables. Consider, for example, the saddle-shaped surface given by the functional operator

$$> \text{z} := (\text{x}, \text{y})- > \text{x}^2 - \text{y}^2; \qquad (5.16)$$

Maple will draw a three-dimensional plot of this function if given the command

$$> \text{plot3d}(\text{z}, -2..2, -2..2); \qquad (5.17)$$

The intersection of this surface with the y=0 plane is an upward-opening parabola which can be viewed by the command

$$> \ \text{plot}(\text{z}(\text{x}, 0), \text{x} = -2..2); \qquad (5.18)$$

whereas the command

$$> \ \text{plot}(\text{z}(0, \text{y}), \text{y} = -2..2); \qquad (5.19)$$

displays the downward-opening parabola resulting from the intersection of the surface with the x=0 plane. Other parallel *slices* through the surface can be plotted by fixing x at other values and plotting. Try, for example,

$$> \ \text{plot}(\text{z}(1, \text{y}), \text{y} = -2..2); \qquad (5.20)$$

Whenever x is fixed at, say, x_o, the function z(x,y) becomes a simple function $z(x_o, y)$ of y whose dependence can be plotted as a curve in two dimensions. The slope of the curve is just the derivative

$$\frac{dz(x_o, \ y)}{dy} = \text{diff}(\text{z}(\text{x}_o, \ \text{y}), \ \text{y}) = -2y. \qquad (5.21)$$

Similarly, the slope of the curve formed with the intersection of the surface with the plane $y = y_o$ is

$$\frac{dz(x, \ y_o)}{dx} = \text{diff}(\text{z}(\text{x}, \ \text{y}_o), \ \text{x}) = 2x. \qquad (5.22)$$

The derivative of a function z(x,y) of several variables with respect to one variable, say x, when all the other variables are held fixed is called the *partial derivative* $\partial z(x, y)/\partial z$. In Maple, the `diff` command is used for partial derivatives. Try

$$> \texttt{diff(z(x,y),x); diff(z(x,y),y);}$$
$$> \texttt{diff(x * sin(y) * cos(u), x);} \tag{5.23}$$

Partial derivatives are useful for calculating *infinitesimal* (vanishingly small) changes in a smooth function of several variables when the coordinates of the argument are displaced by infinitesimal amounts. Consider the change in a function of two variables, and represent the infinitesimal changes by differentials dx, dy, df :

$$
\begin{aligned}
dz(x, \ y) &\equiv z(x + dx, \ y + dy) - z(x, \ y) \\
&= z(x + dx, \ y + dy) - z(x, \ y + dy) + z(x, \ y + dy) - z(x, \ y) \\
&= \frac{\partial z(x, \ y + dy)}{\partial x} dx + \frac{\partial z(x, \ y)}{\partial y} dy \\
&= \frac{\partial z(x, \ y)}{\partial x} dx + \frac{\partial z(x, \ y)}{\partial y} dy. \tag{5.24}
\end{aligned}
$$

In the last step, we expanded $z(x, y + dy)$ about y and ignored terms of second order or higher in the infinitesimals. The extension of this result to the differential of a function $f(x_1, x_2, \ldots, x_n)$ of n variables $x_1, x_2, \ldots, x_n$ is

$$df = \sum_{k=1}^{n} \frac{\partial f}{\partial x_k} dx_k. \tag{5.25}$$

5.5 Errors and Their Propagation

Computers and calculators usually produce answers to 10 decimal places or so, and the layperson not sophisticated in error analysis is often mislead into believing that all the digits produced are meaningful. The experienced scientist, however, knows that if the computer is fed inaccurate data, the results it produces with it will be inaccurate as well. This is epitomized in the famous phrase: *garbage in, garbage out*. However, the effect of erroneous input will differ depending on its role in determining the results. In this section, we will see how elementary calculus can predict how errors will be propagated. We assume that the errors are generally symmetrical and small compared to measured values.

Only rarely do we directly measure the desired quantity. Usually several experimental quantities enter into a functional relation to determine the result. Given the errors (or uncertainties) in the experimental inputs, we want to determine the error (or uncertainty) in the result. These calculations need not be tedious. It is sufficient to estimate errors to one or at most two decimal places. Because the errors are small, products of errors are negligible compared to the errors themselves, and the error propagation in functional relations can be estimated by formulas for differentials. Suppose, for example, that the result R is a function of measured quantities x_k, $k = 1, 2, \ldots : R = R\,(x_1, x_2, \ldots)$. The infinitesimal change in R arising from the infinitesimal shifts dx_j is given as shown in the last result of the previous section:

$$dR = \sum_j \frac{\partial R}{\partial x_j} dx_j, \qquad (5.26)$$

where the sum has been written explicitly.

If the errors δx_j were all correlated so that they all had the same sign, then we could simply use the differential formula with the differentials dx_j replaced by the errors δx_j. The more common situation, however, is that the errors are independent, and consequently, their signs are unrelated to each other. This means that they are as likely to add as to subtract, and that the average value of R from a set of measurements vanishes: $\langle \delta R \rangle = 0$. However, the average of $(\delta R)^2$ is generally not zero. If the errors δx_j are indeed uncorrelated, the cross terms vanish:

$$\langle \delta x_j\, \delta x_k \rangle = \langle \delta x_j \rangle \langle \delta x_k \rangle = 0, \ j \neq k \qquad (5.27)$$

leaving

$$\left\langle (\delta R)^2 \right\rangle = \sum_k \left(\frac{\partial R}{\partial x_k} \right)^2 \left\langle (\delta x_k)^2 \right\rangle. \qquad (5.28)$$

This is the general rule for combining independent errors.

Consider some simple examples.

1. First, let the result R be a sum of independent terms:

$$R = \sum_j a_j \, x_j \tag{5.29}$$

where the coefficients a_j are constants. The average square error in R is the sum of the average square errors in the terms x_j weighted by the squares of the constant coefficients a_j^2:

$$
\begin{aligned}
\left\langle (\delta R)^2 \right\rangle &= \sum_k a_k^2 \left\langle (\delta x_k)^2 \right\rangle \\
&= a_1^2 \left\langle (\delta x_1)^2 \right\rangle + a_2^2 \left\langle (\delta x_2)^2 \right\rangle \cdots
\end{aligned} \tag{5.30}
$$

2. Next, let the result be a product of independent terms. The symbol for a product is based on the Greek letter *pi* in the same way that the sum symbol is based on the Greek *sigma*. Thus

$$R = a \prod_k x_k := a \, x_1 \, x_2 \, x_3 \cdots . \tag{5.31}$$

According to (5.28), the average square error is

$$
\begin{aligned}
(\delta R)^2 &= (a x_2 x_3 ...)^2 \left\langle (\delta x_1)^2 \right\rangle + (a x_1 x_3 \cdots)^2 \left\langle (\delta x_2)^2 \right\rangle + \cdots \\
&= R^2 \left[\left\langle \left(\frac{\delta x_1}{x_1} \right)^2 \right\rangle + \left\langle \left(\frac{\delta x_2}{x_2} \right)^2 \right\rangle + \left\langle \left(\frac{\delta x_3}{x_3} \right)^2 \right\rangle + \cdots \right] \\
&= R^2 \sum_k \left\langle \left(\frac{\delta x_k}{x_k} \right)^2 \right\rangle .
\end{aligned} \tag{5.32}
$$

Exercise 5.2. Suppose the length of a rod is measured in three sections and that the "standard error" of each of the measurements is 2 mm. What would the error for the total length be?
Answer: $2\sqrt{3}$ mm.

Exercise 5.3. The volume of a rectangular box can be determined by multiplying together the lengths of the three sides. If each length measurement is precise only to 0.5%, what is the overall relative precision with which the volume is determined?
Answer: $\frac{1}{2}\sqrt{3}$ %.

Range of X	No. in this range
[25,29] kg	7
(29,31) kg	8
[31,35] kg	5

Table 5.1. Weights of 20 premix bags

5.6   Distributions

Once a sufficient number of independent measurements have been made, it may be useful to plot the results on a *histogram*. For this purpose, we partition the variate space into *bins*. Any possible combination of variate values corresponds to a *position in variate space* and lies in one (and only one) bin. Consider measurements with a single variate, that is, with a one-dimensional variate space. In the histogram, each bin is represented by a bar whose height is proportional to the number of measured values within that bin divided by the "width" of the bin, that is, by the interval of variate values in the bin. The bins are often given the same widths. The bins should all be large enough to hold at least a half dozen measurements if the height of the bar is to be meaningful. As the number of measurements n becomes infinite, the areas of the bins can be reduced to zero while the number of "events" (*i.e.*, measured values) in each bin becomes large. The histogram will then approach a continuous function, namely n times the *population distribution function*, whose value gives the probability that an event occurs in a unit area of variate space at that position.

To make the concepts (the italicized words above) more concrete, consider an example with a single variate, namely the weight X of a pre-mixed bag of sand and lime, and suppose that the values cluster around 30 kg. After measuring 20 bags, we find the following distribution:

There are three bins in this example, given by the ranges of X. The histogram generated by these data consists of three bars: the outer bars have widths of 4 kg, whereas the central one has a width of 2 kg. The bars have heights of 7/4, 4, and 5/4 in units of kg^{-1}. The area of each bar gives the number of measurements with weights in the corresponding bin. The sum of the areas of the bars is equal numerically to the total number of measurements n made (see Fig. 5.2).

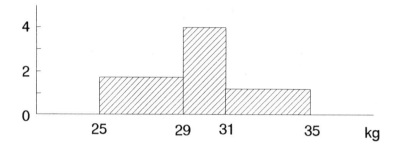

Figure 5.2. The histogram of weights. There are three bins of different widths. The heights of the bars give the average number of bags per kilogram with weights in the weight range of the bin.

If the number of bags measured is increased from 20 to 2000, the variate space can be partitioned into many more bins, and most bins can contain more measured values. We might, for example, choose to have 40 bins with an average of 50 events each. If the heights of the bars are divided by n, the sum of the areas will be unity. Gradually, as more and more bags are measured, the histogram (divided by n) will converge on a continuous function, namely the distribution function $f(x)$ of the population from which the bags were chosen. The integral of the distribution over variate space is the area under the curve, and it will be unity:

$$\int_{-\infty}^{\infty} f(x)dx = 1 \,. \tag{5.33}$$

The integral of the distribution from x_1 to x_2 gives the area under that part of the curve and hence the probability that a randomly chosen bag will have its weight in the range $X \in (x_1, x_2)$.

The average value $\langle x \rangle$, sometimes called the *first moment* of the distribution $f(x)$, is

$$\langle x \rangle = \int_{-\infty}^{\infty} x f(x) \, dx \, . \tag{5.34}$$

More generally, the m-th moment of the distribution is the average value of x^m, given by

$$\langle x^m \rangle = \int_{-\infty}^{\infty} x^m f(x) \, dx \, . \tag{5.35}$$

The variance or square deviation of f is thus

$$\sigma^2 = \left\langle (x - \langle x \rangle)^2 \right\rangle = \int_{-\infty}^{\infty} (x - \langle x \rangle)^2 f(x) \, dx = \left\langle x^2 \right\rangle - \langle x \rangle^2 \, . \tag{5.36}$$

Many distributions are approximately Gaussian in shape:

$$f(x)dx = (2\pi)^{-1/2} \exp(-t^2/2)dt \, , \tag{5.37}$$

where $t = (x - \langle x \rangle)/\sigma$ is the deviation of x from the mean $\langle x \rangle$ in units of the standard deviation σ.   For such distributions, the probability that an event lies within one standard deviation of the mean is

$$\int_{\langle x \rangle - \sigma}^{\langle x \rangle + \sigma} f(x)dx = \int_{-1}^{1} \frac{\exp(-t^2/2)}{\sqrt{2\pi}} dt \approx 0.68 \, . \tag{5.38}$$

See Section 5.9 for further discussion.

The variate space may well be more than one-dimensional. If, for example, there are three independent variates, the space is three-dimensional. Each bin then gives the probability that an event occurs in a given volume of variate space. In this chapter we concentrate on measurements with a single variate. Multivariate problems will be considered in more detail in Chapter 6.

5.7   Samples and Populations

An experiment comprising a set of measurements of a single variate X gives both an average value $\langle X \rangle$ and an rms deviation (see Section 5.2) of the measured values $x^{(m)}$ from $\langle X \rangle$. Another experiment, perhaps on the same equipment under similar conditions but with its own set of measurements of the single variate, generally gives a somewhat different average value and rms deviation. The differences can usually be reduced by increasing the number of measurements in the two experiments. We say that both $\langle X \rangle$ and the rms deviation of the measured values from $\langle X \rangle$ tend to *converge* to fixed values as the number of measurements ("trials") increases.

By comparing results of *experiments* (sets of measurements) rather than of individual measurements, we have jumped to a new level of abstraction. The beauty of statistics is that it can handle this higher level of abstraction as easily as the lower one. In fact, a statistical analysis of the experiments is often more useful than that of the individual measurements.

Statisticians describe an experiment as choosing a *sample* of trials from an underlying *population*. The distribution of values in the sample will usually approximate that in the population, but only in the limit as the number of trials becomes infinite will the two distributions coincide. The average $\langle X \rangle$ of the finite sample is not necessarily that of the population, but its definition (5.1) ensures that the rms deviation of the sample values from $\langle X \rangle$ is minimized.

To prove this assertion, consider the mean square deviation of measured values from an arbitrary value x_0

$$\Delta(x_0) := \frac{1}{n} \sum_{m=1}^{n} (x^{(m)} - x_0)^2 . \tag{5.39}$$

For a given set of measurements $\{x^{(m)}\}$, minimize Δ with respect to x_0:

$$\frac{\partial \Delta(x_0)}{\partial x_0} = \frac{2}{n} \sum_{m=1}^{n} (x_0 - x^{(m)}) = 2 \left(x_0 - \frac{1}{n} \sum_{m=1}^{n} x^{(m)} \right) = 0 \tag{5.40}$$

The solution is seen to be $x_0 = \langle X \rangle$, that is, the mean square deviation is a minimum when taken about the average value $\langle X \rangle$, and of course its square root, the rms deviation, has its minimum at the same place.

The rms deviation of the sample about its own average is thus less than that about the population average, and it is usually less than the standard deviation, that is, less than the rms deviation of the population about its mean. A better estimate of the standard deviation than the rms deviation of the sample is $\sqrt{n/(n-1)}$ times that amount:

$$\sigma_X^* = \sqrt{\sum_{m=1}^{n} \frac{(x^{(m)} - \langle X \rangle)^2}{n-1}} \, . \tag{5.41}$$

The asterisk indicates that this is an estimate. It is what Maple computes with the command sdev in Release 2 and with the command standarddeviation[1] in Release 3.

The argument for this form is as follows. The average value $\langle X \rangle$ of the sample was determined by an experiment with n independent trials, a sample with n *degrees of freedom*, one may say. However, the calculation of the rms deviation uses the $\langle X \rangle$, whose value is a linear combination of the measurements. With $\langle X \rangle$ fixed, the measurements are no longer all independent; instead, they are coupled by the one equation which relates them to $\langle X \rangle$. As a consequence, the number of degrees of independence or freedom is reduced by one, and there are only $n-1$ independent measurement quantities from which to estimate σ_X.

One of the important estimates we can make is the probable error of the sample mean from the mean of the true distribution of the population. The average of a sample (an experiment) is the sum of n independent measurements divided by n:

$$\langle X \rangle = \frac{1}{n} \sum_{m=1}^{n} x^{(m)} \, . \tag{5.42}$$

We can use equation (5.28) for error propagation, where the result R is the average value $\langle X \rangle$ of a sample, and each sample is a set of n measurements. Since for each measurement m of a sample, the partial derivative is

$$\partial \langle X \rangle / \partial x^{(m)} = 1/n \, , \tag{5.43}$$

the square error of $\langle X \rangle$ averaged over all samples of the population is related to the average square error $\langle \delta^2 \rangle$ in each measurement by

$$\left\langle (\delta \langle X \rangle)^2 \right\rangle = \frac{1}{n^2} \sum_{m=1}^{n} \left\langle \delta^2 \right\rangle = \frac{1}{n} \left\langle \delta^2 \right\rangle . \tag{5.44}$$

Thus, the variance of the average value should be smaller than the variance for an individual measurement by the factor $1/n$ where n is the number of trials, and the square root of the variance is

$$\sigma_{\langle X \rangle} = \frac{\sigma_X}{\sqrt{n}} . \tag{5.45}$$

The best estimate of the standard deviation in $\langle X \rangle$, also known as the standard error is therefore[3]

$$\sigma_{\langle X \rangle}^* = \sqrt{\sum_{m=1}^{n} \frac{(x^{(m)} - \langle X \rangle)^2}{n(n-1)}} . \tag{5.46}$$

5.8 Discrete Distributions

From a small sample of measurements, it is often not obvious what the distribution of the underlying population is. Of course, the distribution can be determined by observing a very large number of trials, but it is not always practical to measure such large samples. In this section, we consider cases in which the distribution can be determined *a priori*. We concentrate in particular on situations in which the variate can take only a finite number of distinct values.

When a coin is flipped, there are two possible outcomes: heads or tails. The outcomes are mutually exclusive, so that if the probability for flipping heads is q, that for getting tails is $1 - q$. Furthermore, different flips are independent of each other, so that the probability that in n flips, all the results are heads, is q^n.[4] More generally, the probability

---

[3]The standard error is given in Maple V Release 2 by `serr` . In Release 3, use
> `standarddeviation[1](Xdat)/sqrt(n)`
where `Xdat` is the data list of n values of X. See the worksheets `ws5r2.ms` and `ws5r3.ms` for more information.

[4]This follows from a basic rule of probability theory: the probability that two independent events occur is the product of the probabilities for the individual events. (See also Problem 5 at the end of the chapter.)

that the first m flips give heads and the rest, tails, is $q^m(1-q)^{n-m}$. The probability that any m of the n flips are heads while the rest are tails is this factor times the number of ways of choosing m items from n, namely

$$\mathcal{B}_m^{(n)} = \binom{n}{m} q^m (1-q)^{n-m}. \qquad (5.47)$$

This is the discrete distribution for tossing m heads and $m-n$ tails. It is known as the *binomial distribution* and is applicable whenever the independent trials have only two mutually exclusive outcomes. The binomial theorem shows that the distribution is normalized to unity:

$$\sum_{m=0}^{n} \mathcal{B}_m^{(n)} = \sum_{m=0}^{n} \binom{n}{m} q^m (1-q)^{n-m} = (q+1-q)^n = 1. \qquad (5.48)$$

The average number of heads tossed in n trials is m times the probability that m heads are tossed, summed over all possible m :

$$
\begin{aligned}
\langle m \rangle &= \sum_{m=0}^{n} \mathcal{B}_m^{(n)} m \\
&= qn \sum_{m'=0}^{n-1} \mathcal{B}_{m'}^{(n-1)} \\
&= qn \qquad (5.49)
\end{aligned}
$$

where $m' = m - 1$.

As a more physical example, consider a Geiger counter which is recording the background radiation. Let the average rate R of counts ("clicks") be a few per minute. The counting system works by polling the counter at small time intervals of Δt to see whether the counter has discharged since the previous polling; if it has, a counter is increased by one. The average number of counts during an interval Δt is $q = R\Delta t$. If this number is small compared to unity, it is approximately the probability that at least one count will be detected during Δt. The conditions satisfy the criteria for the binomial distribution: there are two mutually exclusive results of each trail: either the counter tube discharged (counted) or it didn't, and what happens in one interval Δt is independent of whether or not a count was recorded for any other interval. Thus after n time intervals, the probability that m counts were detected is

$$\binom{n}{m} q^m (1-q)^{n-m} . \tag{5.50}$$

The only problem with this formulation is that the probability $q = R\Delta t$ may be a slight overestimate, since two or more counts could in principle occur during Δt but would, in the measurement, be recorded only once. One solution is to make the time interval Δt very small. This will make the probability q for an event in Δt small as well, and it will make the likelihood of two or more events in Δt negligibly small. In the limit of small finite m and

$$
\begin{aligned}
n &\to \infty \\
\Delta t &\to 0 \\
n\Delta t &\equiv \Delta T = const \\
q = R\Delta t &= R\Delta T/n \to 0
\end{aligned}
\tag{5.51}
$$

the binomial distribution becomes (see Problem 2 at the end of the chapter)

$$
\begin{aligned}
\binom{n}{m} q^m (1-q)^{n-m} &\to \frac{(nq)^m}{m!} \exp(-nq) \\
&= \frac{(R\Delta T)^m}{m!} \exp(-R\Delta T) \\
&:= \mathcal{P}_m(R\Delta T),
\end{aligned}
\tag{5.52}
$$

which is known as the *Poisson distribution*. It is the probability that exactly m counts, arriving at the average rate R, are recorded in the time ΔT. For a fixed value of $x \equiv R\Delta T$, it is a distribution in the discrete variable m. As such, it is normalized since the probability that some number counts are recorded in ΔT is

$$\sum_{m=0}^{\infty} \frac{x^m}{m!} \exp(-x) = \exp(x) \exp(-x) = 1 . \tag{5.53}$$

Note that the ratio of the probability for the occurrence of m events to that for $(m-1)$ events is x/m. Thus, m events are more likely to occur than m-1 events as long as $m < x$. Evidently, the probability as a function of the number of events m is largest for the largest m less than x. The average value of m is

$$
\begin{aligned}
\langle m \rangle &= \sum_{m=0}^{\infty} \mathcal{P}_m(x)m \\
&= \sum_{m=0}^{\infty} m \frac{x^m}{m!} \exp(-x) \\
&= \sum_{m=1}^{\infty} \frac{x^m}{(m-1)!} \exp(-x) \\
&= \sum_{n=0}^{\infty} \frac{x^{n+1}}{n!} \exp(-x) \\
&= x,
\end{aligned}
\tag{5.54}
$$

where we replaced the dummy variable m by $n+1$. The variance of m is also x (see Problem 3 at the end of the chapter).

> **Exercise 5.4.** How many counts must be recorded before the result is reliable to within 1%?
> **Answer**: Since the standard deviation is $\sigma = \sqrt{\langle m \rangle}$, the fraction $\sigma/\langle m \rangle = 1/\sqrt{\langle m \rangle} = 0.01$ when the number of counts is $\langle m \rangle = 10000$.

On the other hand, for fixed m, the Poisson distribution $\mathcal{P}_m(x)$ may be considered a distribution in the continuous variable x. The meaning of

$$
\int_0^{R\Delta T} dx\, \mathcal{P}_m(x)
\tag{5.55}
$$

is the probability that m events, arriving at the average rate R, are recorded within the time ΔT. This distribution is also normalized since (see the definition of the GAMMA function in Appendix C)

$$
\int_0^{\infty} dx\, \mathcal{P}_m(x) = 1.
\tag{5.56}
$$

This result simply states the obvious: m events arriving at an average rate R must occur eventually. As the reader can verify, the distribution has its maximum at $x_0 = m$. The average value of x in the distribution is

$$\langle X \rangle \; = \; \int_0^\infty dx \, \mathcal{P}_m(x) \, x \tag{5.57}$$

$$= \; \frac{1}{m!} \int_0^\infty dx \, x^{(m+1)} \exp(-x) \tag{5.58}$$

$$= \; m + 1 \,. \tag{5.59}$$

Exercise 5.5. Show that the variance $\sigma^2 = \left\langle (x - \langle x \rangle)^2 \right\rangle = \langle x^2 \rangle - \langle x \rangle^2$ of the Poisson distribution is equal to $\langle x \rangle$. In other words, $\langle x \rangle = \sigma_x^2 = m + 1$.

5.9 *The Normal (Gaussian) Distribution

The most common distribution is the *normal* or *Gaussian* distribution

$$\frac{1}{\sigma \sqrt{2\pi}} \exp\left[-(x - \langle x \rangle)^2 / 2\sigma^2 \right] \tag{5.60}$$

with an average value $\langle x \rangle$ and a standard deviation σ. Use Maple to plot the Gaussian distribution with a standard deviation of 1 and an average value of $\langle x \rangle = 0$:

$$> \texttt{plot(exp(-x\^{}2/2)/sqrt(2 * Pi), x = -4..4);} \tag{5.61}$$

Note how rapidly the wings of the distribution drop to 0. A wide variety of error distributions can be well approximated by such a function. For large m, for example, the Poisson distribution becomes Gaussian.[5] The fraction of measurements that fall in the range (x_1, x_2) is given by the integral

$$\frac{1}{\sqrt{2\pi}} \quad \int_{x_1}^{x_2} \frac{dx}{\sigma} \exp(-(x - \langle x \rangle)^2 / 2\sigma^2)$$

---

[5] R. L. Plackett, *An Introduction to the Theory of Statistics* (Barnes & Noble, Inc. 1971), p.170–3.

$$= \frac{1}{\sqrt{2\pi}} \int_{(x_1-\langle x \rangle)/\sigma}^{(x_2-\langle x \rangle)/\sigma} dt \, \exp(-t^2/2)$$

$$= \frac{1}{2} \left[\mathrm{erf}\left(\frac{x_2-\langle x \rangle}{\sqrt{2}\sigma}\right) - \mathrm{erf}\left(\frac{x_1-\langle x \rangle}{\sqrt{2}\sigma}\right) \right] \qquad (5.62)$$

where `erf(s)` is the error function:

$$\mathrm{erf}(x/\sqrt{2}) = \frac{1}{\sqrt{2\pi}} \int_{-x}^{x} dt \, \exp(-t^2/2). \qquad (5.63)$$

Exercise 5.6 Show that the error function is monotonically increasing and has the values $\mathrm{erf}(0) = 0$, $\mathrm{erf}(\infty) = -\mathrm{erf}(-\infty) = 1$.

The fraction of measurements within n standard deviations of $\langle x \rangle$ is simply $\mathrm{erf}(n/\sqrt{2})$. Have Maple evaluate this at $n = 1$, 2, and 3 to check the statements at the end of Section 5.2. In particular, verify that results which are quoted as "true 19 times out of 20" are 2σ results, that is, results within 2σ from the average.

5.10   *Model Calculations

We can test much of the above theory with models generated with Maple. To create a data list of 20 normally distributed values, given to four decimal places, define the random number by

$$> \mathtt{N} := \mathtt{RandNormal}(5.0, 1.0, 4) :$$

in Maple V Release 2, and by

$$> \mathtt{N} := \mathtt{random[normald}(5.0, 1.0, 4)] :$$

in Release 3. Then try the command

$$> \mathtt{dat} := [\mathtt{seq}([\mathtt{N}()], \mathtt{i} = 1..20)]; \qquad (5.64)$$

The `RandNormal(av,sd,ndigits)` command in Release 2 and the `random[normald(av, sd, ndigits)]` command in Release 3 generate a pseudo-random number of `ndigits` digits in a normal (Gaussian) distribution centered at `av` with a standard deviation of `sd`. If `ndigits` is not specified, the environment variable Digits is used. A sequence of numbers is pseudo-random when the numbers look random, but in fact follow a repeatable pattern. Now we can compare the average $\langle X \rangle$ of the sample with that of the distribution, namely 5.0:

$$> \texttt{av\_x} := \texttt{mean(dat)}; \tag{5.65}$$

The deviation should be of the order of the standard error $serr = \sigma^*_{\langle X \rangle}$. Following the commands at the end of Section 5.3, we can also compare the rms deviation from the average with the standard deviation of the population, namely 1.0, and with the estimate σ^*_X of the standard deviation. The calculations are easily repeated with larger data sets. Note how the rms deviation approaches the population standard deviation as the number of data is increased.

5.11 *Problems

1. Use the general result at the end of Section 5.5 to derive explicit formulas for the propagation of errors or uncertainties in the following two cases:
$$R = \sum_k a_k (x_k)^{p_k} \tag{5.66}$$
and
$$R = \prod_k (x_k)^{p_k} \tag{5.67}$$
where a_k and p_k are real constants.

2. Use the binomial expansion and the power-series expansion of $\exp(x)$ to show that in the limit $n \to \infty$
$$(1 + x/n)^n \to \exp(x). \tag{5.68}$$

3. Prove that the variance in a counter is equal to the average count:
$$\sigma^2_m = \left\langle (m - \langle m \rangle)^2 \right\rangle = \langle m \rangle . \tag{5.69}$$

4. Assume a normal distribution and find (a) the fraction of measurements which lie within 2.5σ and within 4σ of the average value, and (b) use `fsolve` to find the range of results which are correct 99% of the time.

5. If A and B are events which occur with probabilities $P(A)$ and $P(B)$, respectively, it is a basic rule of probability theory that the probability $P(A \vee B)$—that one *or* the other (or both) occur—plus the probability $P(A \wedge B)$—that both (one *and* the other) occur—is given by the sum of $P(A)$ and $P(B)$:

$$P(A \vee B) + P(A \wedge B) = P(A) + P(B). \qquad (5.70)$$

Furthermore, if the events are *independent*, the probability of both occuring is the product

$$P(A \wedge B) = P(A) * P(B). \qquad (5.71)$$

If two events A and C are *mutually exclusive*, their joint probability is zero: $P(A \wedge C) = 0$. The event *not A*, written $\neg A$, is not only mutually exclusive with A, it is *complementary*:

$$P(A \vee \neg A) = P(A) + P(\neg A) = 1 \qquad (5.72)$$

Use this information to calculate the probability that when 1000 independent measurements of a normally distributed population are measured, at least one of them lies more than 3σ from the average. Your answer should be accurate to at least three significant figures. (Hint: it's probably easiest to consider the negation of this event.)

5.12   Chapter Summary

5.12.1   Concepts:

accuracy and precision

averages

square deviations, rms deviations

standard deviations, variance

data list

partial derivatives

error propagation

histograms, bins

populations and samples

standard deviation of the mean

distributions, moments

discrete distributions: binomial and Poisson

normal (Gaussian) distributions

random and pseudo-random numbers

5.12.2   Maple Commands:

`with(stats)`

`with(describe)` (Release 3)

`mean`

`erf`

`map`

`RandNormal` (Release 2)

`random[normald]` (Release 3)

`standarddeviation` (Release 3)

`serr` (Release 2)

Chapter 6

Curve-Fitting

Scientists are interested in functional relations; they want to know, for example, how the amplitude of some signal changes in time or how the energy changes with position. However, they typically have only a finite set of data, usually values of the function at discrete points of the independent variable(s). To determine values elsewhere, they need to fit a smooth curve to their discrete data. The curve fills the gaps between the data points and can be used to perform numerical integrations. There are two cases to be distinguished: (1) the data points are exact and simply need to be joined by an appropriately smooth curve, and (2) the data points are approximate, perhaps measured values, and a curve of some given form is sought which minimizes their square deviation. Both cases are considered in this chapter.

Consider first fitting smooth curves to exact data. Determining values *between* the given data points is known as *interpolation*. Such results are generally more accurate than when values *outside* the data points are estimated, a procedure known as *extrapolation*. Interpolation and extrapolation are discussed in Sections 6.1 and 6.2. Different conditions can be satisfied by the fitted curve. In the simplest case, the curve merely matches a number of given functional values; four functional values, for example, determine a cubic curve. If the data includes not only the functional values but also the derivatives (slopes) of the function at the discrete points, then the data from two points determine a cubic polynomial. The more points taken together, the higher order is the fitted curve. However, high-order polynomials sometimes

147

give unrealistic oscillations, and as you move from one set of points to the next, discontinuities often appear in the derivatives. Additional conditions ensuring the continuity of one or more derivatives lead to *spline* fits (Section 6.3), which have become quite popular in the past two decades.

On the other hand, if the data points are not exact but only approximate, we should not distort the fitted curve in order to fit them precisely. Instead, it is usually preferable to fit a curve with only a few parameters to the many approximate data points by minimizing the square deviation of the fit from the points. This general procedure is called the *least-squares* method. In the simplest case, all the data are initially (*a priori*) assumed equally reliable and are fit to a straight line. The procedure (Section 6.4) is *linear regression*, and the accuracy of the fit gives information about the reliability of the two parameters (intercept and slope) determined. Of course, more complex curves can also be fit (Section 6.5), and if the data points are known to have varying precision, the fit can be improved by appealing to the principle of maximum likelihood (Section 6.6). The fitting procedure is direct if the curve is linear in the fitting parameters; then there is a unique solution which is found from a single linear matrix equation. However, if the fitting function is not linear in the parameters, an iterative algorithm (Section 6.7) is often needed, and it may not lead to a unique solution.

Similar considerations apply if we seek a fit with several independent variables. If there are two or more independent variables, we will be fitting a surface or hypersurface to the experimental points. However, even if there are many independent variables, there will still be a single weighted sum χ^2 of square deviations of the fitting function from the experimental points. More important for the fitting algorithm is the number of adjustable parameters in the fit. If the function to be fit to the data is a function of a single parameter p, the least-squares method is equivalent to finding the minimum of a curve $\chi^2(p)$, but a function of two or more parameters $p_1, p_2, \ldots$ requires a search for minima of a surface or hypersurface $\chi^2(p_1, p_2, \ldots)$. In general, this is a difficult problem for which there are both local and global search methods. We give an introduction to such methods in Section 6.7.

Material in Chapter 3 on approximations of real functions is important background to sections of Chapter 6 on curve fitting, and the statistical information in Chapter 5 is needed for the discussion in Sections 6 and 7 on multivariate fits to data. Although most of this chapter is not needed in later sections of this text, the work on interpolation in Sections 1 and 2 is useful background for the discussion of numerical integration in Chapter 7.

6.1 *Linear Interpolation

Consider a function $y(x)$ which is known only at certain discrete values of the independent variable x_i. If we plot the known data on a graph of y *versus* x, we obtain a set of points. The simplest way to fill the gap between discrete functional values is to draw a straight line connecting the points. The equation for the line joining the points (x_1, y_1) and (x_2, y_2) is

$$y_{fit}(x) = a + bx, \qquad (6.1)$$

where the parameters a and b are found by solving

$$y_{fit}(x_1) = y_1, \ y_{fit}(x_2) = y_2. \qquad (6.2)$$

The algebra required is pretty easy, but Maple will find the result if you give the commands

```
> yfit := x− > a + b * x;
> eqnset := {yfit1(x1) = y1, yfit(x2) = y2};
> slnset := {a, b};
> solve(eqnset, slnset);
```
(6.3)

It's instructive to express the resulting linear function in the form

$$y_{fit}(x) = \sum_{n=1}^{2} P_n(x)y_n, \qquad (6.4)$$

where the linear polynomials $P_1(x)$ and $P_2(x)$ are given by

$$P_1(x) = \frac{x_2 - x}{x_2 - x_1}, \ P_2(x) = \frac{x - x_1}{x_2 - x_1}. \qquad (6.5)$$

This procedure for approximating intermediate points of a curve is called *linear interpolation*; it is also known as *first-order Lagrange interpolation*. Note that the *coefficient polynomials* $P_n(x)$ have the property

$$P_n(x_m) = \delta_{nm} , \tag{6.6}$$

where δ_{mn} is the Kronecker delta (see Chapter 4). This is a general property that holds for higher-order polynomial interpolation as well: at the tabulated points x_m, the polynomials $P_n(x)$ act as switches which turn on the appropriate function value y_n and turn off all others.

Of course, a curve fitted by straight-line segments is not *smooth*: its slope is usually discontinuous at the data points. As an example, use Maple's `seq` command (an *implicit do loop*) to define a data set of five points of the exponential function. Let h be the step size:

$$> h := 1.0; \tag{6.7}$$

The coordinates of the points can be specified with lists

$$\begin{aligned} &> \text{X} := \; [\text{seq}(\text{h*}(\text{n} - 1), \; \text{n=1..5})]; \\ &> \text{Y} := \; [\text{seq}(\text{exp}(\text{X}[\text{n}]), \; \text{n=1..5})]; \end{aligned} \tag{6.8}$$

A Maple plot of the points

$$\begin{aligned} &> \text{pts} := [\text{seq}[\text{X}[\text{n}], \text{Y}[\text{n}]], \text{n} = 1..5)]; \\ &> \text{pltpt} := \text{plot}(\text{pts}, \text{x} = 0..4, \text{style} = \text{point}); \end{aligned} \tag{6.9}$$

will show the straight-line segments if the default `style=line` is used:

$$> \text{plot}(\text{pts}, \text{x} = 0..4)], \text{style} = \text{line}); \tag{6.10}$$

The point plot can also be drawn together with a plot of a given curve:

$$> \; \text{pltcrv} := \text{plot}(\text{exp}(\text{x}), \text{x} = 0..4); \tag{6.11}$$

with the `display` command in the `plots` package:

$$\begin{aligned} &> \text{with}\,(\text{plots}) : \\ &> \text{display}([\text{pltpt}, \text{pltcrv}]); \end{aligned} \tag{6.12}$$

We can define the linear function which fits the third and fourth points by

```
> yfit1 := x- > Y[3] + (Y[4] - Y[3])/(X[4] - X[3]) * (x - X[3]);
```

Compare this linear approximation to the exponential function graphically:

$$> \; \texttt{plot(\{exp(x), yfit1(x)\}, x = 0..4);} \qquad (6.13)$$

and note that the curve deviates rapidly from its linear approximation in the *extrapolation* region outside the interval x[3] < x < x[4]. Extrapolations are generally subject to higher errors than interpolations and should, as a general rule, be restricted to distances less than about a half of a step size beyond the last point fit.

6.2 *Other Interpolation Procedures

The linear fits discussed above can be generalized to higher-order fits to more than two points. The general form of the N-point polynomial approximation can be written

$$y_{fit}^{(N)}(x) = \sum_{n=1}^{N} P_n^{(N)}(x)\, y_n \qquad (6.14)$$

where the N coefficients $p_n^{(N)}(x)$ are $(N-1)$-degree polynomials in x with the properties

$$P_n^{(N)}(x_m) = \delta_{mn}. \qquad (6.15)$$

For each n, these N conditions $(m = 1..N)$ are sufficient to determine the N coefficients in the $(N-1)$-degree polynomial uniquely. It can be seen that the polynomials[1]

---

[1] As discussed in the last chapter (section 5.5), Π stands for a product, just as Σ indicates a sum. Thus, for example,

$$P_2^{(3)} = \prod_{\substack{m=1 \\ m \neq 2}}^{3} \frac{(x_m - x)}{(x_m - x_n)} = \frac{(x_1 - x)(x_3 - x)}{(x_1 - x_2)(x_3 - x_2)}.$$

$$P_n^{(N)}(x) = \prod_{\substack{m=1 \\ m \neq n}}^{N} \frac{(x_m - x)}{(x_m - x_n)} \tag{6.16}$$

have the desired properties and must therefore be the unique solutions. The resulting polynomial approximation is called *Lagrange interpolation.*

To carry our previous example further, use the Maple concatenation token, namely a period, together with a do loop to define

```
> for n to 5 do x.n := X[n]; y.n := Y[n] od;        (6.17)
```

The third-order Lagrange interpolation formula which passes through the first four points can be written

```
> yfit3 := x - >
y1*(x-x2)*(x-x3)*(x-x4)/((x1-x2)*(x1-x3)*(x1-x4))+
y2*(x-x1)*(x-x3)*(x-x4)/((x2-x1)*(x2-x3)*(x2-x4))+    (6.18)
y3*(x-x1)*(x-x2)*(x-x4)/((x3-x1)*(x3-x2)*(x3-x4))+
y4*(x-x1)*(x-x2)*(x-x3)/((x4-x1)*(x4-x2)*(x4-x3));
```

Compare the linear and cubic approximations graphically with the exponential function:[2]

```
> plot({exp, yfit1, yfit3}, 0..4);               (6.19)
```

Maple has an `interp` command to save you typing (<u>now</u> I tell you!):

```
> ypoly := interp(X, Y, x);                      (6.20)
```

gives the $(N - 1)$-order polynomial in x which passes through the N points $(X[n], Y[n])$, n = 1..N where X and Y are lists. Since in the

---

[2]Alternatively, you can use functional expressions, in place of functional operators, in the plot command:
```
> plot({exp(x),yfit1(x),yfit3(x)},x=0..4);
```

present case, there are five points, the polynomial is a quartic (fourth-degree) polynomial. The simpler quadratic fit to the three middle points is thus

> yquad := interp([X[2], X[3], x[4]], [Y[2], Y[3], Y[4]], x);   (6.21)

Since the expressions ypoly and yquad are *polynomial expressions* and not *functional operators*, we cannot use the functional-operator form of (6.19) unless we first unapply them. However, they can be plotted and compared to the exp(x) function with the command

> plot({exp(x), ypoly, yquad}, x = 0..4);   (6.22)

The related functions are

> yfit2 := unapply(yquad, x);
> yfit4 := unapply(ypoly, x);   (6.23)

Try

> yfit2(1); yfit2(1.3); yfit4(1.3); yfit4(1.5); D(yfit2)(x);   (6.24)

and compare the relative errors

> plot(yfit2/exp-1,-0.1..5);
> plot(yfit4/exp-1,-0.1..5);   (6.25)

The relative error of the quartic fit is shown in Figure 6.1.

Several remarks are in order. Note first of all that the approximations are generally better as interpolations than as extrapolations. Furthermore, the approximations fit to several steps in x are generally better in the central intervals than in the outer ones. Note also the oscillation in the error of the quartic approximation: the quartic approximation is not always better than the linear one. Check the fit over the first interval:

> plot({yfit4,exp},0..1);   (6.26)

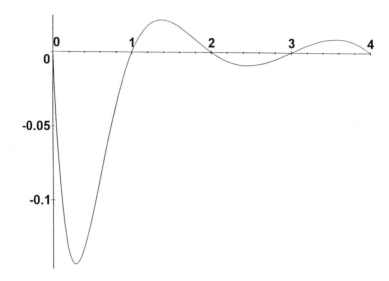

Figure 6.1. The relative error of a quartic fit to points on an exponential curve.

In the neighborhood of $x = 0$, a linear fit from $x = 0$ to $x = 1$ would have given a better fit than the quartic one. The slope of the quartic at $x = 0$ even has the wrong sign. The error is an indication of an instability which can become more severe when higher-order polynomials are forced to fit points on a curve which, like the exponential, is not well described by the polynomial over the range fit.

Other information can be used to derive other interpolation formulas. *Hermite* interpolation, for example, uses both the values of the function, y1,y2,..., and its first derivatives, yp1,yp2,..., at discrete values, x1,x2,..., of its argument. The most common is two-point cubic approximation

$$> \text{yherm} := x- > a + b * x + c * x\char`^2 + d * x\char`^3; \qquad (6.27)$$

whose coefficients can be determined by Maple to fulfill[3]

---

[3]If x1, x2, y1, y2 are still assigned, you will have to "unassign" them:
> x1:='x1';x2:='x2';y1:='y1';y2:='y2';
or, alternatively,

$$> \text{ eqs } := \{\text{yherm}(x1) = y1, \text{yherm}(x2) = y2,$$
$$D(\text{yherm})(x1) = yp1, D(\text{yherm})(x2) = yp2\}; \qquad (6.28)$$

The commands

$$> \text{ solve}(\text{eqs}, \{a, b, c, d\});$$
$$> \text{ assign}("); \qquad (6.29)$$

will find and assign the coefficients needed, whereas the interpolation formula itself is put in reasonable form when we let h be the step size and p*h the distance from x1 to x:

$$> \text{ x2 } := \text{ x1 } + \text{ h};$$
$$> \text{ yherm}(x1 + p * h);$$
$$> \text{ simplify}("); \qquad (6.30)$$
$$> \text{ collect}(", p);$$

The result is the interpolation formula

$$(\ 2 \ y1 - 2 \ y2 + yp1 \ h + yp2 \ h)p\hat{\ }3 + yp1 \ h \ p+ \qquad (6.31)$$
$$(-3 \ y1 + 3 \ y2 - 2 \ yp1 \ h - yp2 \ h)p\hat{\ }2 + y1$$

You can verify that after simplification yherm(x1) = y1, yherm(x2) = y2, D(yherm)(x1) = yp1, and D(yherm)(x2)= yp2.

6.3   *Splines

When fitting a large set of points, one rarely uses polynomials of degree greater than about six. Because of instabilities, higher-order polynomials are not necessarily more accurate, and the numerical precision required is excessive when combining large and small powers of x. Instead of using a single polynomial, one usually uses a segmented polynomial: as the position x moves from one interval to the next, the coefficients for

---

```
> readlib(unassign);
> unassign('x1','x2','y1','y2');
```
where the readlib command locates the library procedure unassign.

the interpolating polynomial generally change. The interpolated curve will be continuous because all of the points are fit by both polynomial segments, but it may not be *smooth*, that is, it may have a discontinuous first derivative. Hermite interpolation gives a continuous slope but will usually have a discontinuous second derivative.

Splines seek to improve on Lagrange-type and Hermite-type interpolations by using conditions of smoothness to help determine the polynomial coefficients. Suppose, for example, that a cubic fit has been obtained for the interval $x_{n-1} < x \le x_n = x_{n-1} + h_{n-1}$ and that we now want to extend the solution to the next interval. Write the cubic approximation for $x_n < x \le x_{n+1} = x_n + h_n$ as a power-series expansion in $p_n := (x - x_n)/(x_{n+1} - x_n) = (x - x_n)/h_n$:

$$y_{spline}(x) = y_n + A_n p_n + \frac{1}{2} B_n p_n^2 + \frac{1}{6} C_n p_n^3 \qquad (6.32)$$

where $y_{spline}(x_m) = y_m$. The conditions of continuity of the first two derivatives of the function at x_n and of the function itself at x_{n+1} give constraints

$$\begin{aligned} A_n &= \left(A_{n-1} + B_{n-1} + \tfrac{1}{2} C_{n-1} \right) (h_n/h_{n-1}) \\ B_n &= \left(B_{n-1} + C_{n-1} \right) (h_n/h_{n-1})^2 \\ y_{n+1} &= y_n + A_n + \tfrac{1}{2} B_n + \tfrac{1}{6} C_n \,, \end{aligned} \qquad (6.33)$$

which are sufficient to determine the coefficients A_n, B_n, C_n for x in $x_n < x < x_{n+1}$ from the values y_n, y_{n+1} and the coefficients in the interval $x_{n-1} < x < x_n$.

In this way, the fit can be splined from one interval to the next. The only remaining question is how to start: how are the coefficients chosen for the first point? This turns out to be a nontrivial question: the wrong choice can lead to instability resulting in wild oscillations in C_n from one interval to the next. A proper spline fit minimizes the fluctuations in C_n.

To find a suitable algorithm, first note that the continuity equations for the function and its second derivative (the third and second equations of (6.33), respectively) can be combined to eliminate A_n and C_n in the interpolation formula (6.32). Putting $q_n := 1 - p_n$, we can write

$$y_{spline}(x) = y_n q_n + y_{n+1} p_n - \frac{p_n q_n}{6}\left[B_n(1+q_n) + B_{n+1}(1+p_n)\left(\frac{h_n}{h_{n+1}}\right)^2\right].$$
$$(6.34)$$

It is therefore sufficient to find the coefficients B_n. For simplicity, we assume the points are evenly spaced: $h_n = h_{n+1}$. First we eliminate C_{n-1} from the first two relations of (6.33). The result is

$$A_n - \frac{1}{2}B_n = A_{n-1} + \frac{1}{2}B_{n-1} \qquad (6.35)$$

which can be combined with the continuity relation at x_n to give

$$y_{n-1} = y_n - A_n + \frac{1}{2}B_n - \frac{1}{6}C_{n-1}. \qquad (6.36)$$

When this is added to y_{n+1}, we find for $1 < n < N$, where N is the total number of points to be fit,

$$
\begin{aligned}
B_n &= y_{n+1} - 2y_n + y_{n-1} - \tfrac{1}{6}(C_n - C_{n-1}) \\
&= y_{n+1} - 2y_n + y_{n-1} - \tfrac{1}{6}(B_{n+1} - 2B_n + B_{n-1}).
\end{aligned}
\qquad (6.37)
$$

The equation can be solved iteratively, starting in the lowest-order approximation with $C_n = C_{n-1}$, so that the last term vanishes and the coefficients at the end points are given by

$$B_1 = 2B_2 - B_3, \ B_N = 2B_{N-1} - B_{N-2}. \qquad (6.38)$$

Usually only a few iterations are required to reach convergence. We continue to assume $C_1 = C_2$ and $C_N = C_{N-1}$ at the end points. The converged values of B_n are used in the interpolation formula (6.34) to give the desired spline curve.

Many modern plotting routines use cubic splines to plot smooth curves. Maple provides a spline routine for calculating splines which can then be plotted. First execute a `readlib` statement so that Maple knows where to find it:

$$> \texttt{readlib(spline)}; \qquad (6.39)$$

The command is then

$$> \texttt{x:='x'; spline(X,Y,x,3);} \qquad (6.40)$$

where X and Y are lists of x and y values, the third parameter instructs Maple to make x the independent variable, and "3" asks for a cubic spline. If the "3" is replaced by "1," a *linear spline* is obtained, which simply connects the points with straight lines. The result of the spline call is short-hand for a segmented polynomial. Fortunately, Maple will make the result into a function procedure when instructed

$$> \texttt{yspline:='spline/makeproc'('',x);} \qquad (6.41)$$

Maple's cubic splines differ somewhat from the ones described above because the conditions on the third derivatives in the end intervals are different. Whereas we let them be equal to the third derivatives in the neighboring interval, Maple has set them to zero to obtain what are called *natural splines*.

Exercise 6.1. Consider the four points of $\exp(x)$ at $x = 0$, 1.0, 2.0, and 3.0. Compare the results of interpolations using (i) 2-point Lagrange, (ii) 5-point Lagrange, (iii) 2-point Hermite, and (iv) the two cubic spline methods to the exact value of the exponential function at $x = 1.7$. For the 2-point methods, use the two points $x_2 = 1$ and $x_3 = 2$.

Solution: Define lists as in (6.8):

$$\begin{array}{ll} > \texttt{h} & := \quad \texttt{1.0;} \\ > \texttt{X} & := \quad [\texttt{seq(h} * (\texttt{n} - 1), \texttt{n} = 1..5)]; \\ > \texttt{Y} & := \quad [\texttt{seq(exp(X[n]), n} = 1..5)]; \\ > \texttt{for n to 3 do YP[n]} & := \quad \texttt{Y[n] od;} \end{array} \qquad (6.42)$$

It's easy to compare the exact value with the Lagrange 2- and 5-point interpolations:

$$\begin{array}{ll} > \texttt{L2pt} & := \quad \texttt{interp([X[2], X[3]], [Y[2], Y[3]], x);} \\ > \texttt{L5pt} & := \quad \texttt{interp(X, Y, x);} \\ > \texttt{x} & := \quad \texttt{1.7; exp(x); L2pt; L5pt;} \end{array} \qquad (6.43)$$

The Hermite result is

```
>p:=(x-X[2])/h;
>(2*Y[2]-2*Y[3]+h*YP[2]+h*YP[3])*p^3+h*YP[2]*p-
  (3*Y[2]-3*Y[3]+2*h*YP[2]+h*YP[3])*p^2+Y[2];
```
$$(6.44)$$

The cubic spline starts with the coefficients

```
>for n from 2 to 4 do
  B[n]:=Y[n+1]-2*Y[n]+Y[n-1] od;
  B[1]:=2*B[2]-B[3]; B[5]:=2*B[4]-B[3];
```
$$(6.45)$$

and after a few iterations of

```
>for n from 2 to 4 do B[n] :=
Y[n+1]-2*Y[n]+Y[n-1]-(B[n+1]-2*B[n]+B[n-1])/6
od;
>B[1] := 2*B[2]-B[3]; B[5] := 2*B[4]-B[3];
```
$$(6.46)$$

the coefficients will have converged and can be used in

```
>q := 1 - p;
>spl:=Y[2]*q+Y[3]*p-p*q*(B[2]*(1+q)+B[3]*(1+p))/6;
```
$$(6.47)$$

For the natural cubic spline, use the three Maple commands given just before this exercise, followed by

```
> yspline(1.7);
```

The results should be

$$
\begin{array}{ll}
\text{exact: } \exp(1.7) = & 5.47395 \\
\text{2-point Lagrange:} & 5.98782 \\
\text{5-point Lagrange:} & 5.54452 \\
\text{Hermite:} & 5.46523 \\
\text{cubic spline:} & 5.50562 \\
\text{natural cubic spline:} & 5.63505
\end{array}
$$

Exercise 6.2. Compare the linear and natural cubic-spline fits to the exact exponential curve graphically.

Solution: compute the linear spline as above but with the "3" replaced by "1":

> x:='x'; spline(X,Y,x,1);

Then define the procedure

> ylin:='spline/makeproc'('',x);

Then use the plot command

> plot({exp, yspline, ylin}, 0..4);   (6.48)

6.4   *Linear Regression

The fits discussed above were for precise data; we forced the curve to pass through the given points. However, it may often happen that the data to be fit are the result of measurements with an inherent random error. Because of the errors, there is no point in fitting each point precisely. Instead, it is more meaningful to fit a large body of such data with a function containing only a few parameters.

Suppose our data are pairs of numbers (x_i, y_i) meaning that the approximate value y_i was measured at the (relatively precise) x_i. We will initially assume that the measured data are all equivalent in that there is no *a priori* reason to assume that some data are more accurate than others. In other words, all measurements in the sample are from a population with a single variance σ_y^2. The simplest functional relationship we might establish between x and y is a linear one:

$$y_{fit}(x) = a + bx. \tag{6.49}$$

The experimental points will generally be scattered about this line. The best values of a and b will be those which give the minimum square deviation of the data from the line. Thus we seek the minimum of

$$\chi^2(a, b) = \frac{1}{\sigma_y^2} \sum_{m=1}^{N} [y_m - y_{fit}(x_m)]^2. \tag{6.50}$$

The normalization factor $1/\sigma_y^2$ ensures that the function χ^2 is dimensionless. As indicated, χ^2 is a function of the parameters a, b; its value is a measure of the accuracy of the fit of the curve $y_{fit}(x)$ to the data points (x_m, y_m). A smaller value of χ^2 indicates a better fit.

To see the functional dependence typical of $\chi^2(a, b)$, consider the following data of the length of a spring in mm as a function of the mass in kg of a weight hanging from the lower end:

$$
\begin{aligned}
&> \text{x} \; := \; \text{'x':y:='y':} \\
&> \text{X} := [0, 1, 2, 3, 4]; \\
&> \text{Y} := [12, 16, 23, 28, 32]; \\
&> \text{pts} := \text{seq}([\text{X[n]}, \text{Y[n]}], \text{n} = 1..5);
\end{aligned}
\tag{6.51}
$$

Remember to invoke the with(stats) command before asking for the mean. In Release 3 or Maple V, also add the command with(describe). Assume a variance of 1 and evaluate χ^2 with the commands

$$
\begin{aligned}
&> \text{yfit} := \text{x}->\text{a}+\text{b}*\text{x}; \\
&> \text{for n to 5 do} \\
&> \text{sqdev[n]} := (\text{Y[n]} - \text{yfit}(\text{X[n]}))^2 \text{od}; \\
&> \text{chisq} := 5 * \text{mean(sqdev)};
\end{aligned}
\tag{6.52}
$$

Next try a three-dimensional plot of chisq as a function of a, b:

$$> \text{plot3d(chisq}, \text{a} = 0..20, \text{b} = 0..20); \tag{6.53}$$

This should reveal chisq as a parabolic surface with a single minimum near $(a, b) = (12, 5)$. You might like to try different domains and views in order to see the minimum more clearly. The minimum is located more precisely by calculating where the slopes $\partial(\chi^2)/\partial a$ and $\partial(\chi^2)/\partial b$ vanish. From (6.50),

$$\frac{\partial(\chi^2)}{\partial a} = -\frac{2}{\sigma_y^2} \sum_{m=1}^{N} [y_m - y_{fit}(x_m)] \, \partial y_{fit}(x_m)/\partial a = 0 \quad (6.54)$$

$$\frac{\partial(\chi^2)}{\partial b} = -\frac{2}{\sigma_y^2} \sum_{m=1}^{N} [y_m - y_{fit}(x_m)] \, \partial y_{fit}(x_m)/\partial b = 0. \quad (6.55)$$

With the given linear form of $y_{fit}(x)$, these conditions give

$$\begin{aligned} a + b \, \langle x \rangle &= \langle y \rangle \\ a \, \langle x \rangle + b \, \langle x^2 \rangle &= \langle xy \rangle \, , \end{aligned} \quad (6.56)$$

which are readily solved to yield

$$a = \frac{\langle x^2 \rangle \, \langle y \rangle - \langle x \rangle \, \langle xy \rangle}{\Delta_x^2}, \quad b = \frac{\langle xy \rangle - \langle x \rangle \, \langle y \rangle}{\Delta_x^2}, \quad (6.57)$$

where $\Delta_x^2 \equiv \langle x^2 \rangle - \langle x \rangle^2$ is the mean square deviation in x.

The statistics jargon for finding the parameters of a fit to the data is *regression*. Maple's `stats` package in Release 2 has built-in procedures for performing linear regressions. The commands[4]

$$\begin{aligned} &> \texttt{dat} := \texttt{array}([[\texttt{x}, \texttt{y}], \texttt{pts}]); \\ &> \texttt{regression}(\texttt{dat}, \texttt{y} = \texttt{a} + \texttt{b} * \texttt{x}); \\ &> \texttt{assign}(\text{"});\end{aligned} \quad (6.58)$$

should return the computed values of `a` and `b`. In Release 3, on the other hand, you can use the `leastsquare` fitting procedure in the `stats` subpackage `fit`:

$$\begin{aligned} &> \texttt{with}(\texttt{fit}) : \\ &> \texttt{leastsquare}[[\texttt{x}, \texttt{y}], \texttt{y} = \texttt{a} + \texttt{b} * \texttt{x}]([\texttt{X}, \texttt{Y}]); \\ &> \texttt{assign}(\text{"});\end{aligned} \quad (6.59)$$

---

[4]The command `linregress` (`dat, y=a+b*x`) should work as well in Release 2. However, as its name implies, it is specialized to linear regressions. Here `Dat` is the statistical matrix, which plays a central role in Release 2 but is absent from Release 3. See the worksheets for Chapter 5 and the Maple help pages for more information.

To plot the results, use the `display` command as in Section 6.1. If the `plots` package is no longer loaded, you must first bring in the table of procedures by issuing the `with(plots)` command. The instructions

$$
\begin{aligned}
&> \text{y} := \text{unapply}(\text{a} + \text{b} * \text{x}, \text{x}); \\
&> \text{plt1} := \text{plot}(\text{pts}, \text{x} = 0..4, \text{style} = \text{points}) : \\
&> \text{plt2} := \text{plot}(\text{y}(\text{x}), \text{x} = 0..4) : \\
&> \text{display}([\text{plt1}, \text{plt2}]);
\end{aligned}
\tag{6.60}
$$

should then plot the linear fit together with the data.

A question naturally arises about the reliability of the evaluated parameters. Because we know the parameters as functions of the measured data, we can use the formula for error propagation derived in the last chapter to determine the reliability. Since the positions x_m are relatively precise (see the beginning of this section), the errors propagate from the N variables y_m to the result a, which can be written

$$
a = \sum_m \left(\frac{\partial a}{\partial y_m} \right) y_m \tag{6.61}
$$

where [see (6.57)]

$$
\frac{\partial a}{\partial y_m} = \frac{\langle x^2 \rangle - \langle x \rangle \, x_m}{N \Delta_x^2}. \tag{6.62}
$$

Since all the measured values y_m are assumed to have the same variance σ_y^2, the error-propagation formula (6.43) gives

$$
\begin{aligned}
\sigma_a^2 &= \sum_m \left(\frac{\partial a}{\partial y_m} \right)^2 \sigma_y^2 \\
&= \sigma_y^2 \frac{\left\langle \left(\langle x^2 \rangle - \langle x \rangle \, x \right)^2 \right\rangle}{N \Delta_x^4}.
\end{aligned}
\tag{6.63}
$$

After expanding the square, we find

$$
\sigma_a^2 = \frac{\sigma_y^2 \, \langle x^2 \rangle}{N \Delta_x^2}. \tag{6.64}
$$

A similar derivation gives the variance for the parameter b :

$$\sigma_b^2 = \frac{\sigma_y^2}{N\Delta_x^2}.$$ (6.65)

Exercise 6.3. Derive the above result (6.65).

If σ_y^2 is not known independently, it can be estimated from the mean square deviation:

$$\sigma_y^{*2} = \frac{N}{N-2}\left\langle [y - y_{fit}(x)]^2 \right\rangle.$$ (6.66)

The factor $N/(N-2)$ is a reflection of the two degrees of freedom lost by fitting the two parameters a and b.

6.5   Other Fits with Linear Parameters

The derivations given in the last section are easily extended to polynomial fitting functions

$$y_{fit}(x) = \sum_{j=0}^{J} a_j x^j.$$ (6.67)

Minimizing χ^2 with respect to a_k we obtain

$$\left\langle x^k y \right\rangle = \left\langle x^k y_{fit}(x) \right\rangle = \sum_j a_j \left\langle x^{j+k} \right\rangle.$$ (6.68)

This gives J+1 equations for the J+1 parameters a_j, $j = 0, 1, 2, \ldots, J$ which can be solved simultaneously to yield all the parameters. The equations can be put in matrix form:

$$\mathbf{Ma} = \mathbf{v}$$ (6.69)

where $\mathbf{M}$ is the square matrix

$$M = \begin{pmatrix} 1 & \langle x \rangle & \langle x^2 \rangle & \langle x^3 \rangle & \cdots & \langle x^J \rangle \\ \langle x \rangle & \langle x^2 \rangle & \langle x^3 \rangle & \cdots & \cdots & \\ \cdots & \cdots & \cdots & \cdots & \cdots & \\ \langle x^J \rangle & \cdots & \cdots & \cdots & \cdots & \langle x^{2J} \rangle \end{pmatrix}$$ (6.70)

and **a** and **v** are the column vectors

$$
\mathbf{a} = \begin{pmatrix} a_0 \\ a_1 \\ a_2 \\ \vdots \\ a_J \end{pmatrix}, \quad \mathbf{v} = \begin{pmatrix} \langle y \rangle \\ \langle xy \rangle \\ \langle x^2 y \rangle \\ \vdots \\ \langle x^J y \rangle \end{pmatrix}. \tag{6.71}
$$

The solution is then

$$
\mathbf{a} = \mathbf{M}^{-1}\mathbf{v}, \tag{6.72}
$$

where $\mathbf{M}^{-1}$ is the inverse matrix to $\mathbf{M}$, that is, $\mathbf{M}\mathbf{M}^{-1} = \mathbf{M}^{-1}\mathbf{M} = \mathbf{1}$, and $\mathbf{1}$ is the unit matrix.

Maple is adept at inverting matrices, although it is more efficient to simply let it solve the matrix equation $\mathbf{M}\mathbf{a} = \mathbf{v}$ directly. Other functions can be fit to the data, and the procedure for finding the parameter values is determined as above as long as the parameters all appear *linearly* in the equation for the curve. The curve $y_{fit}(x)$ is linear in a parameter a iff the partial derivative $\partial y_{fit}/\partial a$ is independent of a. The function

$$
y_{fit}(x) = a + b \exp(cx), \tag{6.73}
$$

for example, is linear in the parameters a, b but NOT in the parameter c. The Maple command `regression` can be used to find the parameters of a fitted curve whenever the curve is linear in all its parameters.

6.6 Data Points with Varying Weights

Some data are more difficult to obtain than others, and it may well be that measurements at some x are less precise that at other values. Perhaps the measuring device is less sensitive at high x or perhaps the experimenter simply took fewer measurements at some values than at others. When different standard deviations σ_{y_m} can be estimated, it is best to replace the above definition of χ^2 with the *weighted sum*

$$\chi^2 = \sum_{m=1}^{N} \frac{[y_m - y_{fit}(x_m)]^2}{\sigma_{y_m}^2} . \tag{6.74}$$

The weighting is reasonable since the smaller the standard deviation, the more precise and hence the more valuable the data. The proof of the weighting follows from the *principle of maximum likelihood,* which states that the best fit is the one for which the measured data would have been most likely. Assuming that the measurements are fully independent and that the measured y_m values are from a population normally distributed about the curve with a standard deviation σ_{y_m}, the probability of the given measurement is proportional to the product of Gaussians (see Section 5.9 and Problem 5 of Chapter 5):

$$\frac{\exp\left\{-\left[\frac{y_1-y_{fit}(x_1)}{2\sigma_{y_1}}\right]^2\right\}}{\sigma_{y_1}} \cdot \frac{\exp\left\{-\left[\frac{y_2-y_{fit}(x_2)}{2\sigma_{y_1}}\right]^2\right\}}{\sigma_{y_2}} \cdots = \frac{\exp\left(-\frac{\chi^2}{2}\right)}{\sigma_{y_1}\sigma_{y_2}\cdots\sigma_{y_N}} \tag{6.75}$$

which is greatest where χ^2 as defined above is a minimum.

6.7   Nonlinear Least-Squares Fits

Often the curve to be fit to the data has nonlinear parameters. Sometimes it may be possible to change the tabulated data so that the fit becomes linear. For example, perhaps the data $\{(x_j, y_j)\}$ represents exponential growth or decay and we wish to find the fractional rate of change. Thus we want to know b in the fit of the data by $y_{fit}(x) = A \exp(bx)$. If instead we use the data $\{(x_j, \ln y_j)\}$, the curve to be fit becomes linear in b: $\ln y_{fit}(x) = a + bx$ where $a = \ln A$. The methods discussed above for linear least-squares fits can then be applied.

There are many other possible curves and surface shapes, however, which cannot be reformulated into linear form. The curve to be fit may contain a sum of exponentials, such as $A_1 \exp(b_1 x) + A_2 \exp(b_2 x)$. The goal is again to minimize the square sum of the weighted deviations

$$\chi^2(A_1, A_2, b_1, b_2) = \sum_j (y_j - y_{fit}(x))^2 / \sigma_j^2 , \tag{6.76}$$

which is equivalent to finding the minimum of the hypersurface χ^2 in the four-dimensional *parameter space* of points $(A_1,\ A_2, b_1, b_2)$. Let $\mathbf{p}$ be the vector from the origin of parameter space to the point $(A_1,\ A_2, b_1, b_2)$. The components of $\mathbf{p}$ are then simply the parameters:

$$p := \begin{pmatrix} p^1 \\ p^2 \\ p^3 \\ p^4 \end{pmatrix} = \begin{pmatrix} A_1 \\ A_2 \\ b_1 \\ b_2 \end{pmatrix}. \tag{6.77}$$

One way to get to the minimum is to slide downhill. An infinitesimal change in the components of $\mathbf{p}$ will give rise to an infinitesimal change in χ^2 :

$$d\chi^2 = \frac{\partial\left(\chi^2\right)}{\partial p^k} dp^k. \tag{6.78}$$

The Einstein summation convention is used here to sum over all four, or more generally, all J parameters. The differentials dp^k are components of an infinitesimal vector $d\mathbf{p}$, and the problem is to find the direction of this vector that gives the largest negative $d(\chi^2)$. It's helpful to define a vector $\mathbf{g}$ whose components are the partial derivatives

$$g_k := \partial(\chi^2)/\partial p^k. \tag{6.79}$$

The infinitesimal change in χ^2 is then given by the dot product of two vectors:

$$d(\chi^2) = g_k dp^k \equiv \mathbf{g} \cdot d\mathbf{p}. \tag{6.80}$$

The dot product of two vectors has its largest negative value when the vectors point in opposite directions. The direction of change in parameter space which gives the largest lowering of the mean square deviation is thus parallel to $-\mathbf{g}$. Making use of the proportionality sign $\propto$ we write $d\mathbf{p} \propto -\mathbf{g}$. The vector $\mathbf{g}$ is called the *gradient* of χ^2, and the gradient operator is often denoted by the *nabla* symbol ∇. The components of the gradient $\mathbf{g}$ are thus

$$\left(\nabla\chi^2\right)_k = g_k = \frac{\partial\left(\chi^2\right)}{\partial p^k}. \tag{6.81}$$

The method of always "stepping" in parameter space in the direction opposite to the gradient $\nabla \chi^2$ is called the *gradient method* or sometimes the *method of steepest descent*. If no higher-order derivatives have been calculated, there is no obvious *a priori* way to choose the step size. However, after evaluations of χ^2 at a couple of steps in the direction of $-\mathbf{g}$, the position of the minimum (or zero) in χ^2 can be estimated. Unfortunately, the method can waste time or even hang up at points which are simultaneously minima along one direction and maxima along another. Such points are called *saddle points* because of the shape of the surface in their vicinity, and they occur with annoying frequency in many multidimensional hypersurfaces. More sophisticated methods test higher derivatives, especially curvatures of the hypersurface, before moving. They can help locate local minima, but even after a minimum is reached, there is no guarantee that it is the lowest minimum; it could be only a local dimple.

The gradient $\mathbf{g}$ must vanish at a minimum, and if it is only a small distance δ from the current position $\mathbf{p}$, it should be given approximately by the first-order Taylor-series expansion. If there is only one parameter p to be fit, the parameter space and the gradient function (6.81) are both one-dimensional. If $p_{\min}$ is the value of the parameter at the minimum in χ^2,

$$0 = g(p_{\min}) \approx g(p) + g'(p)\delta \, , \qquad (6.82)$$

where g' is the derivative of g. The equation is easily solved to estimate the distance δ from the present position to the minimum:

$$\delta = -\frac{g(p)}{g'(p)} \, . \qquad (6.83)$$

Once the current position is moved from p to $p+\delta$, the procedure can be iterated to obtain still better approximations of the minimum position. The method is known as *Newton's method* of finding zeros or roots of a function.[5]

The ideas are similar in a multidimensional parameter space. Let $\mathbf{p}_{\min} = \mathbf{p} + \boldsymbol{\delta}$ be the minimum and expand

---

[5]See Section 8.1 for a further example of Newton's method.

$$\mathbf{g}(\mathbf{p}_{min}) = \mathbf{g}(\mathbf{p} + \boldsymbol{\delta}) = 0 \approx \mathbf{g}(\mathbf{p}) + \frac{\partial \mathbf{g}}{\partial p^k} \delta^k . \qquad (6.84)$$

The components of this equation may be solved together to estimate the displacement $\boldsymbol{\delta}$ required to reach the minimum from any position within the bowl of the minimum. The method is known as the *curvature* or *Gauss-Newton method*. From the domain of parameter space about $\mathbf{p}_{min}$ for which χ^2 is within one unit or so of its minimum, an estimate of the probable errors in the final parameter values is obtained.

If the fitting function $y_{fit}(x)$ is linear in the parameter position $\mathbf{p}$, the curvature method locates the minimum in one jump. Let

$$y_{fit}(x) := f_j(x) p^j , \qquad (6.85)$$

where the functions $f_j(x)$ are independent of the position $\mathbf{p}$ in parameter space, as above, and the Einstein summation convention is used for the sum over all J parameters. The gradient is then

$$
\begin{aligned}
g_j &= \frac{1}{\sigma_y^2} \frac{\partial}{\partial p^j} \sum_{m=1}^{N} \left[y_{fit}(x_m) - y_m \right]^2 \\
&= \frac{2}{\sigma_y^2} \sum_{m=1}^{N} f_j(x_m) \left[y_{fit}(x_m) - y_m \right] ,
\end{aligned}
\qquad (6.86)
$$

which is now also linear in $\mathbf{p}$. As before, N is the number of measured points (x_m, y_m). Derivatives of the gradient are now independent of $\mathbf{p}$:

$$\frac{\partial g_j}{\partial p^k} = \frac{2}{\sigma_y^2} \sum_m f_j(x_m) = \frac{2N}{\sigma_y^2} \langle f_j(x) f_k(x) \rangle . \qquad (6.87)$$

Consequently, all higher-order derivatives of the gradient with respect to the parameters p^k vanish, and the first-order Taylor-series expansion of $g(p + \boldsymbol{\delta})$ used in the curvature method is exact and gives the exact displacement $\boldsymbol{\delta}$ of the minimum $\mathbf{p}_{min}$, no matter how large that displacement is. In particular, we could start at the origin of parameter space: $\mathbf{p} = 0 = \mathbf{p}_{min} - \boldsymbol{\delta}$. The curvature method then gives

$$0 = g_j(0) + \frac{\partial g_j}{\partial p^k} p_{min}^k = \frac{2N}{\sigma_y^2} \left[- \langle f_j(x) y \rangle + \langle f_j(x) f_k(x) \rangle p_{min}^k \right] . \qquad (6.88)$$

Since the curvature method seeks out any position where the gradient vanishes, it can lead you to a maximum or a saddle point of χ^2 as well as to a minimum. It works well if the starting position is within the bowl of a global minimum, but otherwise it needs help. In a refinement of the curvature method, the *Levenberg-Marquardt algorithm* combines the gradient and curvature methods, and a recent modification[6] of the algorithm improves its efficiency and range of applicability by making a more thorough characterization of the local surface. However, even with refinements it can, like any local method, get stuck in a local minimum.

One way to seek out the *global minimum* is to survey the entire surface by sampling its height at a network of grid points. However, for several dimensions, this scanning technique can become prohibitively time-consuming. An approach which requires many fewer function evaluations is the *simplex method*. Maple has a simplex package, but at present it works only on linear functions.

6.8   Problems

1. Equations of motion in physics often involve the second derivative but not the first. Find the coefficients of the cubic interpolation function which is constrained to match the function values $y_1 \equiv y(x_1)$, $y_2 \equiv y(x_2)$ and the second derivatives $w_1 \equiv y''(x_1)$, $w_2 \equiv y_2''(x_2)$. Using the derived formulas and the values of the exponential function and its second derivative at 1.0 and 2.0, compute the interpolated value of the exponential at 1.7 and compare to $\exp(1.7)$. [Answer: the interpolation formula is

$$y_{fit}(x_1 + ph) = qy_1 + py_2 - \frac{pqh^2}{6}\left[(q+1)\,w_1 + (p+1)\,w_2\right] \quad (6.89)$$

and the fit is $y_{fit}(1.7) = 5.42449$; see Exercise (6.1).]

2. Show that the matrix solution for the parameters of a polynomial fit (Section 6.5) gives exactly the results of linear regression when $J = 1$.

---

[6]W. E. Baylis and Atul D. Pradhan, *Comp. Phys. Comm.* **31**, 297-301 (1984).

3. Use Maple commands to fit the data given in the statistical matrix defined in Section 6.4 to a cubic equation. Make a statistical plot showing both the data and the fitted curve. Assume the x-values are exact and calculate the square deviation of the y-data from the curve; then estimate the standard deviation of the data. (How many degrees of freedom have been lost by the fit?)

4. Calculate the gradient of the surface $z = x^2 + y^4$ at the point $(x, y) = (1, 1)$. Does the gradient point directly away from the minimum? You can check your calculation of the gradient by using Maple's procedure `linalg[grad]`:

```
> with(linalg);
> grad(x^2 + y^4, [x, y]);
```

5. Show that when the fitting function $Y(x)$ is a power series as in (6.67), the curvature method gives a result which is equivalent to the matrix equation (6.69). [Caution about notation: there are $J+1$ (and not J) parameters in the Jth-order polynomial fitting function (6.67).]

6.9 Chapter Summary

6.9.1 Concepts

Lagrange interpolation and extrapolation

Hermite interpolation

splines

concatenation

regression

least-squares fits, χ^2

parameter space

principle of maximum likelihood

gradient

gradient or steepest-descent method

Newton's method

saddle points

curvature or Gauss-Newton method

6.9.2   Maple Commands

```
interp
```

```
statplot
```

```
regression
```

```
spline
```

```
'spline/makeproc'
```

```
linalg[grad]
```

```
assign
```

```
readlib
```

```
unassign
```

```
collect
```

Chapter 7

Integration

In Chapter 2, we learned how to differentiate functions using either of two commands. Both

$$> \texttt{diff}(\texttt{f}(\texttt{t}), \texttt{t});$$
$$> \texttt{D}(\texttt{f})(\texttt{t});$$

(7.1)

give the same result, namely the first derivative of the function

$$\texttt{f} := \texttt{t}- > \texttt{f}(\texttt{t})$$

(7.2)

The first form, with `diff`, acts on the functional expression `f(t)`, whereas the second, with D, gives a new mapping. A derivative gives the rate of change of a function with respect to its argument. Thus, the time derivative of the position x of a particle is its time-rate of change, in other words the velocity, and the derivative of the velocity is the acceleration. Similarly the time-rate of change of the energy is the power, and so on. Graphically, the derivative is a new function whose value is the slope of the original one.

We have seen many applications of derivatives in this text. In Chapter 3, we used derivatives of a function at a given point to determine ever better approximations to the function in neighborhoods of the point; the approximations were Taylor-series expansions about the point. In Chapter 4, we extended the concept of differentiation from real-valued functions to vector-valued ones, such as position or velocity, which may be viewed mathematically as arrays of real-valued functions (the vector components). The time derivative of a vector-valued function (for example, the velocity) is another vector function (the acceleration, in the example).

In Chapters 5 and 6, we extended differentiation to functions of more than one variable. In a plot of a function of two variables, for example, the function gives the height of a surface above the position given by the values of its two variables. At times it may be convenient to think of the two variables as components of a vector in a two-dimensional variable space. The value of the function provides the third dimension of the three-dimensional space in which the surface lives. An extension to functions of N variables ($N > 2$) lets the function be interpreted as a *hypersurface* in a space of $N + 1$ dimensions: the "height" of the hypersurface above a position in the N-dimensional variable space gives its value. In partial differentiation, all but one of the variables are held fixed, and the derivative is taken with respect to the remaining one. The result is the slope of the line formed by the intersection of the hypersurface with a two-dimensional plane containing the axis of the selected variable and the function axis (usually the vertical axis).

Integration, as the inverse operation to differentiation, is every bit as important in the physical sciences as differentiation. It forms the subject of the present chapter. We start in Section 7.1 with indefinite integration and look there at Maple's ability to perform integration analytically. Definite integration and the interpretation of the integral as an area is discussed in Section 7.2. Numerical integration methods, based on simple interpolation formulas developed in Sections 6.1 and 6.2, are introduced in Section 7.3, with refinements and examples from Maple given in Section 7.4. The solutions of some second-order differential equations are discussed in Section 7.5, with particular emphasis on the harmonic oscillator. Phase-space plots are introduced as a way of visualizing the solutions in Section 7.7, and they are discussed further in Sections 8 and 9 of Chapter 8. The chapter concludes with numerical examples of an orbital problem in two dimensions.

7.1   *Indefinite Integration

Integration is the inverse operation of differentiation. Thus, the derivative of the integral of a function is the original function. (The integral of the derivative of a function is the original function plus an arbitrary constant, as we discuss below.)

The *indefinite integral* of $f(x)$ with respect to x is written either in the form

$$F(x) = \int f(x)\,dx \qquad (7.3)$$

or the fully equivalent form

$$F(x) = \int dx\, f(x)\,. \qquad (7.4)$$

This indefinite integral is defined to be any function of x whose derivative is $f(x)$:

$$f(x) = \frac{d}{dx} F(x) = \frac{d}{dx} \int dx\, f(x)\,. \qquad (7.5)$$

Because the derivative of a constant is zero, if $F(x)$ is an indefinite integral of a given function $f(x)$, then so is $F(x) + c$, where c is any constant. Of course, that's why it is called an *indefinite* integral: it is arbitrary to within an additive constant. As a result, the integral of the derivative of a function may differ from the original function by such a constant:

$$\int dx \frac{df(x)}{dx} = f(x) + c \qquad (7.6)$$

where c is an arbitrary constant. Often authors do not bother to write the arbitrary constant c, but it should not be forgotten.

Like differentiation, integration is a *linear operation*: if $f(x)$ and $g(x)$ are functions of x and if λ and μ are constants, then

$$\int dx\, [\lambda f(x) + \mu g(x)] = \lambda \int dx\, f(x) + \mu \int dx\, g(x)\,. \qquad (7.7)$$

The Maple command for indefinite integration has the form

$$> \ \texttt{int(f(x),x);} \qquad (7.8)$$

It omits the integration constants. Try the following Maple commands:

$$> \texttt{int}(\texttt{x\^{}3} + \texttt{x\^{}5}, \texttt{x});$$
$$> \texttt{int}(\texttt{sin}(\texttt{x}), \texttt{x});$$
$$> \texttt{int}(\texttt{ln}(\texttt{x}), \texttt{x}); \tag{7.9}$$
$$> \texttt{int}(\texttt{x} * \texttt{arcsin}(3 * \texttt{x}), \texttt{x});$$

Maple does a nice job of replacing tables of integrals.

An indefinite integral (7.4) is the solution of a first-order *differential equation* (7.5). For some physical problems, it may be more natural to ask Maple to solve the differential equation than to integrate. For example, consider an object moving vertically in a uniform gravitational field, experiencing a constant acceleration $-g$. To find its velocity as a function of time, we can ask Maple to solve the differential equation $dv/dt = -g$. This is accomplished with Maple's $\texttt{dsolve}$ command:

$$> \texttt{dsolve}(\texttt{diff}(\texttt{v}(\texttt{t}), \texttt{t}) = -\texttt{g}, \texttt{v}(\texttt{t})); \tag{7.10}$$

which writes out the integration constants explicitly. Equivalently, we could integrate $-\int g\, dt$:

$$> \texttt{v} := \texttt{int}(-\texttt{g}, \texttt{t}); \tag{7.11}$$

A second integration should give the position as a function of time:

$$> \texttt{y} := \texttt{int}(\texttt{v}, \texttt{t}); \tag{7.12}$$

The only problem with using $\texttt{int}$ in this case is that the integration constants are easily forgotten. Not only must we add an integration constant to (7.12), we should also have included the integration constant from (7.11) in the integrand of (7.12). It may be safer to solve directly the *second-order differential equation* $d^2y(t)/dt^2 = -g$:

$$> \texttt{y} := \texttt{'y'};$$
$$> \texttt{dsolve}(\texttt{diff}(\texttt{y}(\texttt{t}), \texttt{t\$2}) = -\texttt{g}, \texttt{y}(\texttt{t})); \tag{7.13}$$

The form

$$> \texttt{dsolve}((\texttt{D@@2})(\texttt{y})(\texttt{t}) = -\texttt{g}, \texttt{y}(\texttt{t})); \tag{7.14}$$

is equivalent. The two integration constants in the solution now appear explicitly. They are usually determined by initial conditions, for example, by the position and velocity of the object at $t = 0$. Such conditions can be added to the first parameter in the dsolve command. Thus, to find the vertical position as a function of time of an object starting at ground level with an upward velocity v0, we can enter

$$
\begin{aligned}
&> \texttt{de} := (\texttt{D@@2})(\texttt{y})(\texttt{t}) = -\texttt{g} \\
&> \texttt{ic} := \texttt{y(0)} = 0, \texttt{D(y)(0)} = \texttt{v0} \\
&> \texttt{dsolve}(\{\texttt{de}, \texttt{ic}\}, \texttt{y(t)});
\end{aligned}
\tag{7.15}
$$

The result can be used to find, say, how long it takes for the object to return to the ground:

$$
\begin{aligned}
&> \ \texttt{assign(");} \\
&> \ \texttt{solve}(\texttt{y(t)} = 0, \texttt{t});
\end{aligned}
\tag{7.16}
$$

There are two solutions, corresponding to the time at which the object left the ground and that at which it returned. Call the latter solution t1. The height reached is found by the subs command, which substitutes the relation of the first parameter into the expression given in the second:

$$
> \ \texttt{h} \ := \ \texttt{subs}(\texttt{t} = 0.5 * \texttt{t1}, \texttt{y(t)});
\tag{7.17}
$$

You might have been tempted to ask Maple to evaluate y(0.5 * t1). However, that won't work because Maple understands y(t) as an expression, not as a mapping or a functional operator.[1] To check that the velocity vanishes at 0.5*t1, try

$$
> \ \texttt{subs}(\texttt{t} = 0.5 * \texttt{t1}, \texttt{diff(y(t), t)});
\tag{7.18}
$$

Exercise 7.1. Evaluate the velocity as the object strikes the ground. Verify that its square is

$$
v_0^2 = 2gh \, .
\tag{7.19}
$$

---

[1]To obtain a functional operator, we can apply the unapply command (see Sections 3.7 and 6.2).

7.2   *Definite Integration

As discussed in Chapter 3, the derivative $f(x) = dF(x)/dx$ of a curve $F(x)$ is its *slope*. The integral $F(x)$ of $f(x)$ also has a simple interpretation. Since

$$dF(x) = F(x + dx) - F(x) = f(x)dx, \qquad (7.20)$$

the increase in the integral as the argument is increased from x to $x+dx$ is the *area* under the curve $f(x)$ between x and $x+dx$. More generally, the difference $F(x_2) - F(x_1)$ in the indefinite integrals is called the *definite integral* from x_1 to x_2 and represents the area under the curve $f(x)$ between x_1 and x_2. It is written

$$F(x_2) - F(x_1) = \int_{x_1}^{x_2} dx\, f(x). \qquad (7.21)$$

Note that the integration constant has been eliminated by the subtraction.

Maple will compute the definite integral if the domain of the integration variable is given. Thus, the integrals

$$\int_0^5 dx\, x^2, \ \int_0^\infty dx\, (\sin x)/x, \ \int_{-\infty}^{+\infty} dx\, \exp(-x^2) \qquad (7.22)$$

are computed by the commands

```
>  int(x^2, x = 0..5);
>  int(sin(x)/x,  x = 0..infinity);                    (7.23)
>  int(exp(-x^2), x = -infinity..infinity);
```

Exercise 7.1. Find the area enclosed by the sine curve $\sin(x)$ and the x-axis between $x = 0$ and $x = \pi$. This is the same as the area of a triangle formed by the x-axis, the tangent to the sine curve at $x = 0$, and what vertical line?

Answer: 2. The line $x = 2$.

As an application, consider a charge q initially at rest but subject to a pulse

$$E(t) = E_0 t(1-t)\sin(10t),\ 0 < t < 1 \tag{7.24}$$

of oscillating electric field. We want to find its velocity at the end of the pulse, that is, at $t = 1$. Since its acceleration is

$$a = qE/m, \tag{7.25}$$

we find in units of qE_0/m,

$$> \mathtt{v := int(t * (1 - t) * sin(10 * t), t = 0..1);} \tag{7.26}$$

Note that if Maple can't integrate an expression, it simply returns the integral. It may still be possible to find a power series for the integral. Try, for example,

$$\begin{aligned}&> \quad \mathtt{int(sin(x\char`\^5), x);}\\&> \quad \mathtt{series(", x, 17);}\end{aligned} \tag{7.27}$$

Definite integrals which cannot be expressed analytically can often be approximated numerically:

$$\begin{aligned}&> \quad \mathtt{int(sin(x\char`\^(5/3)), x = 0..4);}\\&> \quad \mathtt{evalf(");}\end{aligned} \tag{7.28}$$

Maple should return the value 0.91637.

Definite integrals are often used to determine *average values* of a function over an interval. Thus, the average value of $f(x)$ over the interval $a < x < b$ is

$$\langle f(x) \rangle_{a<x<b} = \frac{1}{b-a} \int_a^b f(x)\, dx. \tag{7.29}$$

Exercise 7.2. Find the average value of $\sin^2(x)$ over the domain $0 < x < \pi/2$. Extend your result to domains $m\pi/2 < x < n\pi/2$, where m, n are any integers, and to the entire real x-axis. Repeat for the function $\cos^2(x)$.

Answer: 1/2 for all cases.

Exercise 7.3. Interpret $f(x)$ in (7.29) as the slope of a smooth, bounded function $F(x)$. Show that the average value of the slope of $F(x)$ over an interval $a < x < b$ is the same as the slope of the straight line (the secant) cutting $F(x)$ at $x = a$ and $x = b$.

7.3   *Simple Numerical Integration Methods

One way to approximate a definite integral $\int y(x)dx$ numerically is to evaluate the integrand $y(x)$ at a finite number of points, to interpolate between the points with polynomials in x, and to integrate the interpolation functions. The linear interpolation between points $y_0 = y(x_0)$ and $y_1 = y(x_1)$ can be written

$$y_{int}^{(1)}(x) = qy_0 + py_1 \tag{7.30}$$

where $p = (x - x_0)/h = 1 - q$ and the step size is $h = (x_1 - x_0)$. Integration of $y_{int}^{(1)}(x)$ from x_0 to x_1 gives

$$\int_{x_0}^{x_1} dx\, y_{int}^{(1)}(x) = h \int_0^1 dp[(1-p)y_0 + py_1]$$
$$= \frac{h}{2}(y_0 + y_1). \tag{7.31}$$

This is the formula for the area of a trapezoid, and as an approximation to $\int_0^1 dx\, y(x)$, it is known as the *trapezoidal rule* (see Fig. 7.1). An integral over many points is approximated by summing the trapezoidal contributions between each pair of points. If all the points are separated by the same *step size h*, the sum can be written

$$\int_{x_0}^{x_N} y(x)dx \approx h \left[\frac{y_0}{2} + \sum_{n=1}^{N-1} y_n + \frac{y_N}{2} \right]. \tag{7.32}$$

Higher-order polynomial interpolations generally give more accurate results. Thus the quadratic interpolation connecting three equally spaced points x_0, $x_1 = x_0 + h$, and $x_2 = x_1 + h$

$$y_{int}^{(2)}(x) = \frac{y_0}{2}q(1+q) + y_1p(1+q) + \frac{y_2}{2}p(p-1) \tag{7.33}$$

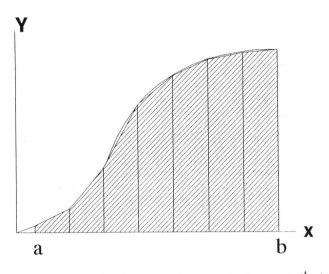

Figure 7.1. The trapezoidal rule approximates the integral $\int_a^b y(x)dx$ by summing the areas of trapezoids.

is integrated to give

$$\int_{x_0}^{x_2} y_{int}^{(2)}(x)dx = \frac{h}{3}(y_0 + 4y_1 + y_2). \tag{7.34}$$

Simpson's rule is the approximation to an integral obtained by summing the above expression over many sets of three points:

$$\int_{x_0}^{x_N} y(x)dx \approx \frac{h}{3}(y_0 + 4y_1 + 2y_2 + 4y_3 + \cdots + 4y_{N-1} + y_N). \tag{7.35}$$

Note the coefficients of all but the outermost terms alternate between $2h/3$ and $4h/3$. Furthermore, the approximation requires an odd number of points and therefore an even number of steps.

The trapezoidal and Simpson methods yield approximate values of the integral which should approach the exact result as the number of equally spaced intervals becomes large. Similar algorithms can be written down for unequally spaced function values, that is, with different step sizes h for each interval. However, they require more computation, and the algorithms with equal spacing are usually preferred. Sometimes the integration variable is changed so that a rapidly varying part of the function is spread over more points.

In applications, it is often important to know how large the error is likely to be and how the error varies with the step size h. An estimate of the error for a single integration step can be made by taking the lowest-order term which was ignored by the algorithm and integrating it. Thus, in the trapezoidal method, the quadratic variation of $y(x)$ was ignored. Since the linear approximation to $y(x)$ is exact at $p = 0$ and $p = 1$, the quadratic correction must vanish there. Furthermore, its second derivative at $p = 0$ must be $y''(x_0)$. These conditions are sufficient to show that in the domain $x_0 < x < x_1$, the lowest-order correction to the linear interpolation must have the form $h^2 y''(x_0)p(p-1)/2$ and the integral of this term from x_0 to $x_1 = x_0 + h$ gives an estimate of the lowest-order correction for the single step:

$$ h \int_0^1 dp\, y''(x_0)h^2 p(p-1)/2 = -\frac{h^3}{12} y''(x_0). \qquad (7.36) $$

The error in the trapezoidal approximation is thus minus this amount. The errors in adjacent intervals are usually *not* independent; unfortunately, they tend to have the same sign so that the total error for the integral as a whole is best approximated by the sum

$$ \epsilon \approx \frac{(b-a)h^2}{12} y''(\xi) \qquad (7.37) $$

where ξ is the point $a < \xi < b$ where the second derivative takes its average value[2]

$$ y''(\xi) = \frac{1}{N} \sum_{n=0}^{N-1} y''(x_n) \approx \frac{1}{(b-a)} \int_a^b dx\, y''(x). \qquad (7.38) $$

For example, if the number of steps is doubled, so that the step size h is halved, the error should drop to about a quarter of its previous value. Let's check this out on the integral

---

[2]It is a fundamental theorem of calculus (*Mean-Value theorem*) that there always exists such a point as long as $y''(x)$ is continuous over the interval $a \le x \le b$.

$$\int_0^1 \exp(x)dx = \exp(1) - \exp(0) = e - 1. \qquad (7.39)$$

The Maple commands

```
> approx1 := evalf((1 + E)/2); approx2 := ";
> for n to 3 do approx1 := approx1 + exp(0.25 * n) od;
> for n to 7 do approx2 := approx2 + exp(0.125 * n) od;
> approx1 := approx1 * 0.25; approx2 := approx2 * 0.125;
```
$$(7.40)$$

calculate trapezoidal approximations to the integral with step sizes of 0.25 and 0.125. The error `approx1-evalf(E-1)` should be roughly four times larger than that with half the step size, `approx2-evalf(E-1)`. Indeed, the former is 0.00894 whereas the latter is 0.00224.

Exercise 7.4. Show that these errors agree well with (7.37).

A similar analysis of Simpson's method shows that the third-order error vanishes, leaving the fourth-order term

$$\epsilon \approx \frac{2(b-a)h^4}{3 \cdot 5!} D^4 y(\xi). \qquad (7.41)$$

Exercise 7.5. Derive the above result.

Hint: Represent the integral as a sum of $(N-1)/2$ three-point integrals of the form

$$h \int_{-1}^1 dp \, y_{int}^{(2)}(x) \qquad (7.42)$$

and note that the error $y_{int}^{(2)}(x) - y(x)$ can be expanded in the polynomial $p(1 - p^2)h^3[D^3 y(x_0)/3! + hpD^4 y(x_0)/4!]$.

7.4 Other Numerical Techniques

We can use any interpolation method to derive an algorithm for integration. Generally the more accurate the interpolation scheme is, the more accurate the integration algorithm will be. An algorithm with a small fourth-order error can thus be obtained from 4-point (cubic) Lagrange interpolation (see Section 6.2). For four evenly spaced points $\{x_{-1}, x_0, x_1, x_2\}$ with $x_n = x_{n-1} + h$, the interpolation scheme can be written in the matrix form

$$y_{int}^{(3)}(x) = \tag{7.43}$$

$$\left[(0,1,0,0) + \frac{p}{6}(-2,-3,6,-1) + \frac{p^2}{2}(1,-2,1,0) + \frac{p^3}{6}(-1,3,-3,1) \right] \mathbf{y}$$

where $\mathbf{y}$ is the column vector

$$\mathbf{y} = \begin{pmatrix} y_{-1} \\ y_0 \\ y_1 \\ y_2 \end{pmatrix} \tag{7.44}$$

and $p = (x - x_0)/h$. This expression is easily integrated from x_0 to x_1:

$$h \int_0^1 dp\, y_{int}^{(3)}(x) = \frac{h}{24}(-1,13,13,-1)\mathbf{y}. \tag{7.45}$$

To approximate the integral $\int_a^b dx\, y(x)$ we simply sum such contributions over all sets of 4 adjacent points. The result requires not only the $N + 1$ values $y_n, 0 \le n \le N$ from $a = x_0$ to $b = x_N$, but also the points $y_{-1} = y(a - h)$ and $y_{N+1} = y(b + h)$ just outside the range of integration. It can be written as a correction to the trapezoidal approximation:

$$\int_a^b y_{int}^{(3)}(x)dx = \frac{h}{2}\left(y_0 + 2\sum_{n=1}^{N-1} y_n + y_N \right) + \frac{h}{24}(y_1 - y_{-1} + y_{N-1} - y_{N+1}).$$
$$\tag{7.46}$$

The error is of fourth order and proportional to the fourth derivative $D^{(4)}y(\xi)$ at some intermediate point ξ:

$$\epsilon \approx -\frac{11(b-a)h^4}{6!}D^{(4)}y(\xi). \tag{7.47}$$

The error is the same order as for Simpson's rule (its magnitude is 2.75 times larger). It is surprising that the relatively simple correction to the trapezoidal approximation can improve its accuracy by two orders. A similar algorithm requiring only points within the range of integration can be derived by integrating the first and last 4-point sets over the first and last steps, respectively, and the error, still fourth order, is essentially the same size:

$$\int_a^b y_{int}'^{(3)}(x)dx = \frac{h}{2}\left(y_0 + 2\sum_{n=1}^{N-1} y_n + y_N\right) + \frac{h}{24}[] \qquad (7.48)$$

where

$$[] = [7(y_1 + y_{N-1}) + y_3 + y_{N-3} - 4(y_0 + y_N + y_2 + y_{N-2})]. \qquad (7.49)$$

The error is

$$\epsilon \approx \frac{h^4}{6!}\{19h[D^{(4)}y(\xi_a) + D^{(4)}y(\xi_b)] - 11(b-a)D^{(4)}y(\xi)\}. \qquad (7.50)$$

Here, ξ_a, ξ_b are points in the first and last intervals of the domain of integration, respectively. For reference, the correction to the trapezoidal rule is also given using the 6-point method for open intervals; it has a small sixth-order error:

$$\frac{h}{1440}[11(y_{-2} - y_2 + y_{N+2} - y_{N-2}) - 82(y_{-1} - y_1 + y_{N+1} - y_{N-1})]. \qquad (7.51)$$

Other interpolation schemes such as ones using derivatives can also be used (see Problem 3). Sometimes you may need to approximate integrals at points x_n which are not evenly spaced. It may be, for example, that the values x_n, y_n are given by some fixed procedure or experiment; you must then start with the more general interpolation formulas. On the other hand, the evaluation of $y(x)$ may simply be very expensive, and you may want to squeeze as much accuracy as possible from just a few evaluations. Then a *Gauss-type* integration may be useful. The integral is approximated by N terms at N points

$$\int_a^b y(x)dx \approx \sum_{n=1}^{N} w_n y(x_n), \qquad\qquad (7.52)$$

where the *weights* w_n and the points x_n are usually chosen so as to satisfy the $2N$ conditions that any polynomial up to (but not including) order $2N$, when multiplied by *density function* $\rho(x)$, integrates exactly. Tables of weights and positions have been tabulated for several different density functions.[3]

7.5   *Numerical Solutions of ODE's

Many of the equations of physical science take the form of second-order ordinary differential equations (ODE's). The equation (7.13) for the position of a falling body is one such equation; another is the equation of motion of a simple-harmonic oscillator, discussed in Section 3.2. Our previous work with such equations involved general analytic solutions; here we want to concentrate on numerical solutions.

To gain some insight into how such equations can be solved, consider the differential equation

$$\frac{d^2 y(x)}{dx^2} = g(x)y(x). \qquad\qquad (7.53)$$

This equation is said to be *linear* in the function $y(x)$ because every term contains the function to the first power.[4] This has the form of the equation of motion of a simple-harmonic oscillator when $g(x)$ is negative. To start a numerical integration, two initial or boundary conditions must be specified. Usually either $\{y(x_0), y'(x_0)\}$ or $\{y(x_{-1}), y(x_0)\}$ are given. The differential equation is then used to determine, to some order of approximation, the values at a neighboring point.

---

[3]See, for example, M. Abramowitz and I. A. Stegun, *Handbook of Mathematical Functions with Formulas, Graphs, and Mathematical Tables*, U. S. Government Printing Office, Washington, D. C. (1964).

[4]Physicists often call an equation linear in $y(x)$ if it consists of terms in both the zeroth power and the first power of $y(x)$. To distinguish the case above, an equation with $y(x)$ appearing in every term to first order is said to be *homogeneous* as well as linear.

The algorithm can be based on a Taylor series expansion about the starting point x_0:

$$y(x) = y_0 + Dy_0(x-x_0) + \frac{1}{2}D^2y_0(x-x_0)^2 + \frac{1}{6}D^3y_0(x-x_0)^3 + \cdots, \quad (7.54)$$

where D is the derivative operator and a subscript n indicates that the function or its derivative is evaluated at the point $x_n = x_0 + nh$ with step size h. The sum $y(x_1) + y(x_{-1})$ contains only even derivatives:

$$y_1 + y_{-1} = 2y_0 + h^2 D^2 y_0 + \frac{h^4}{12}D^4 y_0 + O(h^6). \quad (7.55)$$

A similar relation holds for the second derivative:

$$D^2 y_1 + D^2 y_{-1} = 2D^2 y_0 + h^2 D^4 y_0 + O(h^4). \quad (7.56)$$

The fourth-derivative term can be eliminated in these two equations, and with the help of the original equation for the second derivative, one finds

$$y_1 = \left(1 - g_1\frac{h^2}{12}\right)^{-1}\left[2y_0 - y_{-1} + \frac{h^2}{12}(10g_0 y_0 + g_{-1}y_{-1})\right] + O(h^6). \quad (7.57)$$

Thus, from the known values at x_{-1} and x_0, the function $y(x)$ can be computed at x_1 with a sixth-order error. By iterating the procedure over and over, the entire function can be found. This is the *Numerov method* for solving such second-order differential equations. It has been extended to coupled second-order equations and to equations with inhomogeneous terms.[5]

Maple, in order to solve differential equations numerically, rewrites the equations as coupled first-order equations, so that the equation $D^2 y = gy$ becomes

$$\begin{aligned} Dw(x) &= gy(x) \\ Dy(x) &= w(x), \end{aligned} \quad (7.58)$$

for example. Then a fairly common algorithm known as the *Runge-Kutta method* with a 5th-order error is brought to bear.

---

[5] W. E. Baylis and S. J. Peel, *Comp. Phys. Commun.* **25**, 7–19 (1982).

Only part of the problem in finding a suitable algorithm is to reduce the size of the error term in each step. More important is to ensure that the solutions are *stable*, that is, that any error appearing does not grow unacceptably fast as the solution is propagated. The stability of a solution depends both on the algorithm and on the equation to be solved.

An analysis of stability is beyond the scope of this course, but it should be mentioned that the Numerov and Runge-Kutta methods are relatively stable for most linear equations encountered in physical problems.

Maple will try to compute a numerical solution if the initial conditions are given and if the option numeric is specified. For example, to find the height of a ball thrown upward with an initial velocity of 30 m/s, we could type

$$
\begin{aligned}
&> \text{y} := 'y'; \\
&> \text{eq} := (D@@2)(y)(t) = -9.8; \\
&> \text{ic} := y(0) = 0, D(y)(0) = 30; \\
&> \text{height} := \text{dsolve}(\{eq, ic\}, y(t), \text{numeric});
\end{aligned} \tag{7.59}
$$

Maple responds with a procedure reference and we can then ask for specific values:

$$
> \ \text{height}(1.0); \ \text{height}(2.0); \ \text{height}(5.0); \ \text{height}(10.0); \tag{7.60}
$$

The Maple procedure checks the error and adjusts the step size to maintain the given relative and absolute errors. While the values of the corresponding parameters _RELERR, _ABSERR (see Maple's help information about numeric solutions of dsolve) can be set directly, it is probably safer, at least to begin with, to use the default value 10^(3-Digits) and to control the accuracy by adjusting the parameter Digits (see Section 2.5).

Maple has a convenient command for plotting solutions to ODE's. It's called odeplot and resides in the plots package:

$$
\begin{aligned}
&> \text{with(plots)} : \\
&> \text{odeplot(height}, [t, y(t)], t = 0..10);
\end{aligned}
$$

7.6 *Symbolic solutions of ODE's

Maple is able to find symbolic solutions for many simple ODE's. The equation of motion of a damped harmonic oscillator is an important example. The motion is given by the differential equation

$$\frac{d^2x}{dt^2} + \Gamma\frac{dx}{dt} + \omega^2 x = 0 \tag{7.61}$$

where Γ is the damping constant and ω is the natural angular frequency. Let Maple solve for the motion of the damped oscillator when it receives an impulse at $t = 0$:

```
> ODE := (D@@2)(x)(t) + Gamma * D(x)(t) + omega^2 * x(t) = 0;
> ic := x(0) = 0, D(x)(0) = 1;
> dsolve({ODE, ic}, x(t));
```
$$\tag{7.62}$$

Since the right-hand side (rhs) of the result is the function we want, assign

$$> \text{ X } := \text{ rhs(")}; \tag{7.63}$$

Maple's result can be expressed in the form

$$x(t) = e^{-\Gamma t/2} \sin(\nu t) / \nu \tag{7.64}$$

where

$$\nu = \sqrt{\omega^2 - \Gamma^2/4}\,. \tag{7.65}$$

It is usually convenient to cast results in terms of dimensionless variables. Instead of writing results as a function of t, which has dimensions of time, we can view them as a function of ωt, which is dimensionless. Alternatively, we can consider ωt to be the time in units of ω^{-1}. Similarly, the ratio Γ/ω is just the damping constant Γ in units of ω, and $\omega/\omega = 1$ is the angular frequency itself measured in units of ω. Thus, setting $\omega = 1$ in equation (7.61) does not restrict the physics of the equation; it simply means we are measuring the time in units of ω^{-1} and Γ in units of ω. In addition, we try plotting the results for a case of strong damping, with a constant $\Gamma = 1$:

> omega := 1;
> X1 := subs(Gamma = 1, X) ;   (7.66)
> plot(X1, t = 0..10);

The subs command has replaced Γ by 1 in the solution X. Now we can also plot the velocity:

> V := diff(X1, t);
> plot(V, t = 0..10);   (7.67)

With a *parametric plot* we can also look at the trajectory of V vs. X:

> plot([X1, V, t = 0..10]);   (7.68)

7.7   *Phase Space

The first two plots in the above example show the position and velocity of the damped harmonic oscillator as a function of time. You have probably seen such plots many times before. The third plot shows another way to represent the data: it is a *phase-space* plot of the velocity *versus* position, also known as a *phase-space trajectory* of the motion. We can start a system at any point in phase space, and the phase-space trajectory through that point shows how the system will evolve in time. A collection of such trajectories describes the dynamical behavior of the system by showing its evolution from any starting point; it is called a *phase portrait* of the system (see Problem 8).

Physicists usually take phase space to be a plot of *momentum vs.* position so that an area in a phase-space diagram has units of distance times momentum, and hence of angular momentum. However, since the momentum is just mass times velocity, only a scale factor of m distinguishes momentum *vs.* position plots from velocity *vs.* position ones.

Phase-space plots are popular in studies of chaos. The coordinates of a starting point in such a plot give both of the initial conditions needed to solve a second-order differential equation. The plot of non-chaotic periodic motion is a single closed trajectory in phase space. If you solve for the motion of an undamped harmonic oscillator by redoing the example in Section 7.6 with $\Gamma = 0$, the phase space orbit should be an ellipse. With damping, the orbit decays: it circles in a clockwise sense (why?) and gradually approaches the stable fixed point at the origin. The origin is a *fixed point* because if the oscillator ever has the values $(x, v) = (0, 0)$, it will stay fixed at that point. The point is *stable* because the oscillator will return there if it is displaced a small distance away. We will encounter more phase-space plots in the next chapter.

> **Exercise 7.6.** Redo the plots of Section 7.6 with different values of the damping constant. Try plots of *critical damping* $\Gamma = 2$ as well as *sub-critical* ($\Gamma < 2$) and *super-critical* ($\Gamma > 2$) damping. In the critical damping case you may need to redefine X1 in order to keep Maple from choking on a division-by-zero error (see Problem 6). Alternatively, approximate the case of critical damping by setting $\Gamma = 1.9999$. When $\Gamma < 0$, the origin is still a fixed point, but it is now unstable rather than stable. Try $\Gamma = -0.5$. Can you figure out how the phase-space trajectory indicates whether a fixed point is stable or not? Be careful to note the direction of motion on the trajectory.

> **Exercise 7.7.** The response of a damped oscillator to an impulse varies with damping constant. To view the variation, try the 3-D plot
>
> > plot3d(X, t = 0..10, Gamma = .25..2.5, axes = boxed); (7.69)
>
> Try viewing the plot from several different vantage points (see Figure 7.2.)

7.8 *Coupled Differential Equations

Differential equations encountered in the physical sciences are often more complex than second-order ODE's. We conclude this chapter with the example of coupled second-order equations which arise in the study of planetary motion. The motion of an asteroid in the gravitational

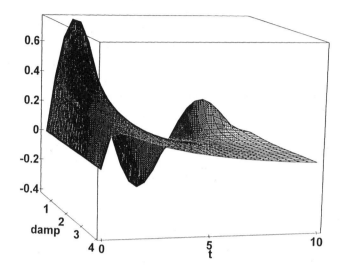

Figure 7.2. The response of a damped oscillator to an impulse depends on the damping constant.

field of the sun is determined (nonrelativistically) by Newton's second law of motion

$$m\ddot{\mathbf{r}} = \mathbf{F} \tag{7.70}$$

where the force $\mathbf{F}$ has the form

$$\mathbf{F} = -\frac{km}{r^2}\hat{\mathbf{r}} \tag{7.71}$$

and where m is the mass of the asteroid, $\mathbf{r} \equiv r\,\hat{\mathbf{r}}$ is its position with respect to the sun, $\ddot{\mathbf{r}} \equiv d^2\mathbf{r}/dt^2$, and $k = GM$ is a constant. The equation of motion is thus a second-order differential equation for the vector position $\mathbf{r}$. It can be expanded into a second-order differential equation for each component of $\mathbf{r}$, and the initial conditions amount to specifying the initial position and velocity $\{\mathbf{r}(0), \dot{\mathbf{r}}(0)\}$. The two vectors $\mathbf{r}$ and $\dot{\mathbf{r}}$ determine a plane, and since the acceleration is directed along $-\hat{\mathbf{r}}$, it lies in the same plane. Thus all the motion occurs in this plane, the invariant *orbital plane* of the motion.

It is convenient to choose axes with their origin at the sun and oriented with $\mathbf{e}_3$ normal to the plane. The position and its time derivatives

can then be expressed as vectors in the two-dimensional space spanned
by the basis $\{e_1, e_2\}$. We write

$$\mathbf{r} = x\mathbf{e}_1 + y\mathbf{e}_2. \tag{7.72}$$

The orbital motion is then determined by two coupled second-order
differential equations:

```
> r := t -> sqrt(x(t)^2 + y(t)^2);
> eqs := {(D@@2)(x)(t)  =  -k*x(t)/r(t)^3,          (7.73)
           (D@@2)(y)(t)  =  -k*y(t)/r(t)^3};
```

where we used $r^2 = x^2 + y^2$. Next we need the initial conditions. To
relate the basis to the orientation of the orbit, let the asteroid be located
on the x-axis when at *perigee* (*perihelion* in this case), that is, when
it is closest to the sun. We can take time of perigee to be $t = 0$ and
measure lengths in units of the distance from the sun to the asteroid
at perigee: $\mathbf{r}(0) = \mathbf{e}_1$.

```
> ic := {  x(0) = 1, D(x)(0) = 0,                    (7.74)
            y(0) = 0, D(y)(0) = sqrt(k*(1 + e))};
```

which should give an orbit of eccentricity e. We need only specify the
constants and then solve numerically:

```
> k := 1; e := .25;
> F := dsolve(eqs union ic, {x(t), y(t)}, numeric);   (7.75)
```

Note that to specify the set of equations we want dsolve to solve, we
have taken the union of two sets: the differential equation set and the
set of initial conditions. The call to dsolve with the numeric option
produces a procedure with which we can calculate data points in the
form of sets. For example, the commands

```
> F(0); F(.5); F(1);                                  (7.76)
```

may produce

$$\{x(t) = 1., y(t) = 0, t = 0\}$$
$$\{t = .5000000000, y(t) = .5366241500, x(t) = .8793556441\}$$
$$\{t = 1., x(t) = .5617564500, y(t) = .9568463745\}$$

$$(7.77)$$

The precise form may vary, but the important point is that the order of the elements in sets is random and generally changes from one data point to the next. The set format of the data points is convenient for use with the subs command. Thus to find the second derivatives at $t = 1$, we can use

$$> \text{subs}(F(1), \text{eqs}); \qquad (7.78)$$

To build a data sequence of x, y points on the orbit for plotting, we start with an empty (NULL) sequence and employ

```
> data := NULL;
> for tt from 0 by 0.25 to 10 do                          (7.79)
    data := data, subs(F(tt), [x(t), y(t)])od :
```

Each time the subs command is executed, it adds a two-member list [x(t),y(t)] to the data sequence data, in which x(t),y(t) are replaced by their values in the set F(tt). To plot the result, it is sufficient to type

$$> \text{plot}([\text{data}]); \qquad (7.80)$$

Exercise 7.8. Plot the orbits for several values of the constants k, e. You may wish to decrease the value of Digits to speed up the process, but make sure that sufficient accuracy is maintained to produce closed orbits. You may also wish to tabulate more data points in order to have a better picture of the orbits. Next, try some of the same plots with the force $\mathbf{F} = -km\hat{\mathbf{r}}/r^3$ in place of the usual gravitational law. Are the orbits still closed? Convince yourself that there is no problem with the accuracy of the results.

7.9   Problems

1. Suppose you are in a closed railway car with no view of the outside. With the aid of your watch and pocket accelerometer, you

know that since starting from rest at $t = 0$, the train has had an acceleration given in $m/\min^2$ by

$$a = t(10 - t)(20 - t) \tag{7.81}$$

where t is the time in minutes.

(a) When does the car stop?

(b) How far did it travel before coming to rest?

2. The technology of the 22nd century may succeed at boring a hole straight through the center of the earth.

(a) Refer back to Section 3.1 and show that the acceleration of an object in such a hole is

$$4\pi R^2 g \int_0^r \rho(r')r'^2 dr' / (Mr^2) \tag{7.82}$$

where r is the distance from the center of the earth, $\rho(r)$ is its density at distance r from the center, R is the radius of the earth, and M is its mass:

$$M = 4\pi \int_0^R \rho(r')r'^2 dr' . \tag{7.83}$$

(b) Recall the relation between the square of the velocity, the acceleration, and the distance travelled:[6]

$$\frac{d(v^2)}{dy} = 2v\frac{dv}{dy} = 2\frac{dy}{dt}\frac{dv}{dy} = 2\frac{dv}{dt} . \tag{7.84}$$

Find the velocity of an object dropped down the hole from the surface as it passes through the center of the earth under the approximation $\rho = $ constant.

---

[6]If this relation is multiplied by $m/2$, it relates the kinetic energy gain to the applied force: $mv^2/2 = \int F dy$.

(c) Repeat the calculation for a density profile of the form

$$\rho(r) = \rho_c[1 + (2r/R)^2]^{-1}, \qquad (7.85)$$

where ρ_c is a constant representing the density of the earth at its center.

3. The 2-point interpolation formula using both the values $y_0 = y(x_0)$, $y_1 = y(x_1)$ and the second derivatives $w_0 = y''(w_0)$, $w_1 = y''(w_1)$ (see Problem 1 in Chapter 6) can be written

$$Y(x) \approx (1-p)y_0 + py_1 + \frac{h^2}{6}p(1-p)[(p-2)w_0 - (p+1)w_1] \quad (7.86)$$

where $h = x_1 - x_0$ and $p = (x - x_0)/h$. Use this interpolation scheme to derive an approximate integration formula for use when both the function and its second derivative are known at equally spaced points.

4. What is the value of height(10) for the tossed ball in Section 7.5? What is the significance of the sign of the answer?

5. Consider the damped harmonic oscillator problem in Section 7.6. What is the phase-space position of the oscillator at time t = 10 when Γ = 1? If you were to estimate the angular frequency of the oscillator from the zero crossings, what would that frequency be (in units of the natural frequency ω)?

6. Show that in the limit of critical damping, when $\Gamma = 2\omega$, the response of the oscillator to an impulse is proportional to $t \exp(-\Gamma t)$.

7. Solve for the motion of a damped oscillator which is released from rest with a displacement of unity from equilibrium. Plot the motion with plot3d as a function of time (in units of ω^{-1}) and damping constant strength (in units of ω).

8. Create a phase-space portrait for the damped oscillator with $\Gamma = \omega$ by superimposing phase-space trajectories for several starting conditions. This can be accomplished by combining separate parametric plots [see (7.68)] with the display command in

the `plots` package, or by invoking the `phaseportrait` command in the `DEtools` package. In particular, the following commands should work:

$$
\begin{aligned}
&> \texttt{with(DEtools);}\\
&> \texttt{ics} := \{[0, 0, .25], [0, 0, .5], [0, 0, .75], [0, 0, 1]\};\\
&> \texttt{phaseportrait}([\texttt{y}, -\texttt{y} - \texttt{x}], [\texttt{x}, \texttt{y}], 0..8, \texttt{ics},\\
&\quad \texttt{stepsize} = .2, \texttt{title} = \text{`Damped Oscillator'});
\end{aligned}
\tag{7.87}
$$

The second-order differential equation (7.61) has been replaced by two coupled first-order equations (for $\Gamma = \omega$)

$$
\begin{aligned}
dx/dt &= y\\
dy/dt &= -y - x
\end{aligned}
\tag{7.88}
$$

as indicated in the first argument of `phaseportrait`. Read the help file about `phaseportrait` to understand the other parameters. Compare to the phaseportrait obtained when $\Gamma = \omega/2$.

7.10 Chapter Summary

7.10.1 Concepts

indefinite and definite integration

differential equations

trapezoidal and Simpson's rules

higher-order integration formulas

Gauss-type integration

Numerov and Runge-Kutta methods

damped oscillator $(\ddot{x} + \Gamma\dot{x} + \omega^2 x = 0)$

critical damping $(\Gamma = 2\omega)$

phase space $(\dot{x} \ vs. \ x \ \text{or} \ m\dot{x} \ vs. \ x)$

phase-space trajectories

phase portraits

orbital motion

invariant orbital plane $(\mathbf{r}, \mathbf{v})$

7.10.2 Maple Commands

```
int
```

```
dsolve, dsolve/numeric
```

```
evalc
```

```
rhs
```

```
odeplot
```

```
NULL
```

```
subs
```

```
op
```

```
union
```

```
for ... from ... by ... to
```

```
do ... od
```

```
with(DEtools); phaseportrait
```

Chapter 8

Complex Numbers and Fractals

With complex numbers, mathematics becomes richer, and its applications to the physical sciences, more marvelous. The numbers with an *imaginary* term, proportional to $i = \sqrt{-1}$, have practical applications far beyond what the famous mathematician Descartes (1596–1650) could have imagined when he rejected them.[1]

This chapter reviews the properties of complex numbers and explores their relation to two-dimensional vectors and their applications in solutions to differential equations, particularly for uniform circular motion and oscillation. There is even a brief introduction to the nonlinear motion of a pendulum. The stunningly beautiful images of the freeware program `Fractint` are also discussed with an introduction to fractals, in particular to the Julia and Mandelbrot sets. Chapter 4 gives important background material on vectors, and Chapter 7 provides essential information for Sections 8 and 9. This chapter, in turn, is required preparation for Chapter 9.

---

[1]René Descartes coined the term "imaginary" to distinguish elements proportional to $\sqrt{-1}$ from real numbers. Many mathematicians of his day, who were still struggling with the new concept of negative numbers, were not willing to accept imaginary numbers into serious mathematics. See M. Klein, *Mathematical Thought from Ancient to Modern Times* (Oxford U., 1972), vol. I, p. 253.

8.1   *Algebraic Roots

If a function $f(x)$ crosses the x-axis $(f = 0)$ at x_r, it is said to *have a (real) root* at that point. A polynomial function of odd degree must always have at least one root, since it is continuous and its leading term asymptotically has opposite signs when $x \to \infty$ and when $x \to -\infty$. Similarly, a polynomial of even degree can have zero or more roots. A polynomial may have as many roots as its degree; thus, for example, a quadratic function (a polynomial of second degree) may have 0 or 2 roots, and a cubic function (a polynomial of third degree) may have one or three roots.

Let Maple draw the plots

$$> \text{plot}(x\char`^3 - x, x = -2..2);$$
$$> \text{plot}(x\char`^3 + x, x = -2..2); \tag{8.1}$$

which should cross the x-axis three times and once, respectively. Roots are given immediately if the function is factored into linear terms. Thus, as Maple will quickly confirm when asked to

$$> \text{factor}(x\char`^3 - x); \tag{8.2}$$

the cubic $x^3 - x$ can be factored into

$$x^3 - x = x(x - 1)(x + 1) \tag{8.3}$$

and can be zero if any one of the factors vanishes:

$$x = 0, \quad x - 1 = 0, \quad x + 1 = 0 . \tag{8.4}$$

These are the roots of the cubic. However, the cubic $x^3 + x$ factors into $x(1 + x^2)$, and the quadratic factor $(1 + x^2)$ has no real roots.

If the function is not a polynomial or if the linear factorization is not obvious, real roots can often be found numerically, for example, by Newton's method, which iterates the solution to a first-order Taylor-series approximation for the root:

$$f(x + \delta) = 0 \approx f(x) + D(f)(x) * \delta . \tag{8.5}$$

In other words, it iterates the replacement

$$x \to x + \delta = x - \frac{f(x)}{D(f)(x)} . \tag{8.6}$$

Exercise 8.1. Use Newton's method to find the square root of 2, that is, the root of $x^2 - 2$.

Solution: Start with a guessed value of x and iterate

$$x \to x - \frac{f(x)}{D(f)(x)} = x - \frac{x^2 - 2}{2x} = \frac{x}{2} + \frac{1}{x}. \qquad (8.7)$$

The Maple commands

$$> x := 1; \text{for } n \text{ to 5 do } x := x/2.0 + 1.0/x \text{ od}; \qquad (8.8)$$

should produce the sequence 1, 1.5, 1.416667, ... A more general form of the commands is

```
> x := 1.0; del := 1; f := x- > x^2 - 2;
> while abs(del) > 10^(-6)                       (8.9)
    do del := -f(x)/D(f)(x); x := x + del od;
```

Although the quadratic factor $x^2 + 1$ which appears in the cubic polynomial $x^3 + x = x(x^2 + 1)$ has no real roots, if we define the *imaginary* number $i = \sqrt{-1}$, roots to the algebraic equation $z^2 + 1 = 0$ can be written

$$z = \pm i. \qquad (8.10)$$

The variable z is said to be a *complex* variable. In general, a complex number has both real and imaginary terms. One writes

$$z = x + iy \qquad (8.11)$$

where x is the real term and iy (where y is real) is the imaginary one:[2]

$$x = \Re z \equiv \langle z \rangle_{\Re}, \quad iy = i\Im z \equiv \langle z \rangle_{\Im}. \qquad (8.12)$$

---

[2]By a rather unfortunate convention, the real number y is usually called the *imaginary part* of $z = x + iy$. It would have been more accurate to call iy the imaginary part, but conventions are hard to change. We will refer to y as the *imaginary component* of z and use the notation $iy = \langle z \rangle_{\Im}$ to distinguish the imaginary term iy from the imaginary component $y = \Im z$.

Every complex number can be represented by a pair of real numbers (x, y) which may be taken as the coordinates of a point (or the components of a vector) in a two-dimensional space called the *complex plane*. The distance of that point from the origin (a Euclidean metric is assumed) is called the *absolute value* $|z|$ of z:

$$|z| = |x + iy| = \sqrt{x^2 + y^2} . \tag{8.13}$$

Exercise 8.2. Here's a simple "join-the-points" and "draw-the-picture" puzzle, but with complex numbers: $1, 1+2i, 1.5(1+i), 3i, 1.5(-1+i), -1+2i, -1$. Join them all in the sequence given and see if the result is recognizable.

The roots of $z^2 + 1$ are imaginary. They may be visualized by plotting the $z^2 + 1$ as a function of x, y. Unfortunately the function $z^2 + 1$ is itself complex:

$$z^2 + 1 = (x + i y)^2 + 1 = (x^2 - y^2 + 1) + 2 i x y. \tag{8.14}$$

It is zero where both its real and imaginary parts vanish. To visualize its complex roots, we can plot its real and imaginary components as functions of x, y:[3]

```
> plot3d({0, x^2 - y^2 + 1}, x = -3..3, y = -3..3);
> plot3d({0, 2 * x * y}, x = -3..3, y = -3..3);
```
$$\tag{8.15}$$

In both cases, the $z = 0$ plane has been plotted with the given function. You may wish to rotate the plot of the imaginary part in order to see it clearly. Both surfaces are *saddle shaped*, that is, they curve upward along one direction and downward along another, but they are rotated from each other by 45°. Both vanish ("intersect the complex plane") at $y = \pm 1$ on the *imaginary axis* ($x = 0$). Since both the real and imaginary parts of $z^2 + 1$ vanish at its roots, so does its absolute value,

---

[3]Remember, you may have to 'unassign' x:

```
> x :='x';
```

and furthermore, its roots are the only points where $|z^2 + 1|$ is equal to zero. To see this graphically, try

```
> z := x + I * y;
> plot3d({0, abs(z^2 + 1)}, x = -3..3, y = -3..3,          (8.16)
    orientation = [14, 86], axes = framed);
```

You should see a bowl with two feet which touch the xy plane at the points $(x, y) = (0, \pm 1)$.

8.2 Vector Spaces

Complex numbers may be thought of as real two-dimensional vectors: they satisfy the requirements of a vector space:

(i) the sum $z = z_1 + z_2$ of any two vectors $z_1 = x_1 + i y_1$, $z_2 = x_2 + i y_2$ is another vector $z = x + i y$ where $x = x_1 + x_2$, $y = y_1 + y_2$;

(ii) the product of a vector with real numbers (scalars) λ, μ is another vector, and the products obey the rules

$$\lambda (z_1 + z_2) = \lambda z_1 + \lambda z_2$$
$$(\lambda + \mu) z = \lambda z + \mu z \qquad (8.17)$$
$$(\lambda \mu) z = \lambda(\mu z);$$

(iii) the product of the unit scalar 1 with a vector is just the vector itself: $1z = z$.

To make the association with vectors more concrete, think of the complex number $z = x + i y$ as the vector from the origin $(0, 0)$ to the point (x, y) in the complex plane. The complex number $-z = -x - i y$ is then a vector from the origin pointing in the opposite direction.

Complex numbers are usually written as the *sum* of real and imaginary terms, but this is not meant to imply that the real and imaginary parts are ever thoroughly combined. On the contrary, the real and imaginary parts of a complex number can always be separated, and an equality between two complex numbers implies both that the real components are equal and, separately, that the imaginary components are equal:

$$z_1 = x_1 + i y_1, \ z_2 = x_2 + i y_2$$
$$z_1 = z_2 \Rightarrow x_1 = x_2 \ \& \ y_1 = y_2. \qquad (8.18)$$

A special case of this relation was used in Section 8.1:

$$z = x + iy = 0 \Rightarrow x = 0, \ y = 0. \tag{8.19}$$

The plus sign in a complex number such as $z = 3 + 5i$ is more appropriately thought of as the "and" in the statement "3 apples and 5 cookies" than as an actual addition operation which would blend the two together as in apple crunch. However, the use of an addition sign does indicate that complex numbers are more than just two-dimensional vectors. We explore this further below.

8.3   Powers and Functions

The major difference between complex numbers and two-dimensional vectors is that an associative multiplication of complex numbers is well defined:

$$\begin{aligned} z_1 z_2 &= (x_1 x_2 - y_1 y_2) + i(x_1 y_2 + y_1 x_2), \\ (z_1 z_2) z_3 &= z_1 (z_2 z_3) =: z_1 z_2 z_3, \end{aligned} \tag{8.20}$$

where we noted that real numbers commute with the imaginary i: $xi = ix$. In particular, the power of an imaginary number is real or imaginary when the power is even or odd:

$$(iy)^n = \begin{cases} 1, & n = 0 \\ iy, & n = 1 \\ -y^2, & n = 2 \\ -iy^3, & n = 3 \\ y^4, & n = 4 \\ \cdots & \end{cases} . \tag{8.21}$$

Therefore, any analytic function of a real variable, that is, a function which can be expanded in a power series, can be defined as well for an imaginary variable. For example, the exponential function becomes

$$\begin{aligned} e^{i\eta} &= 1 + i\eta - \tfrac{\eta^2}{2!} - i\tfrac{\eta^3}{3!} + \tfrac{\eta^4}{4!} + \cdots . \\ &= \left(1 - \tfrac{\eta^2}{2!} + \tfrac{\eta^4}{4!} + \cdots\right) + i\left(\eta - \tfrac{\eta^3}{3!} + \tfrac{\eta^5}{5!} - \cdots\right) \\ &= \cos\eta + i\sin\eta. \end{aligned} \tag{8.22}$$

The function of an imaginary variable is generally complex: its even part is real whereas its odd part is imaginary:

$$\Re f(iy) \equiv \langle f(iy) \rangle_\Re = \tfrac{1}{2}\left[f(iy) + f(-iy) \right]$$

$$i\Im f(iy) \equiv \langle f(iy) \rangle_\Im = \tfrac{1}{2}\left[f(iy) - f(-iy) \right] .$$

$$(8.23)$$

Exercise 8.3. From the definitions $\cosh z = \left[\exp(z) + \exp(-z) \right]/2$ and $\sinh z = \left[\exp(z) - \exp(-z) \right]/2$ of the hyperbolic trig functions, prove the relations

$$\cosh(i\alpha) = \cos\alpha$$
$$\sinh(i\alpha) = i\sin\alpha .$$

$$(8.24)$$

8.4 *Wessel-Argand Diagrams

The splitting $\exp(i\eta) = \cos\eta + i\sin\eta$ of the complex exponential into real and imaginary terms is known as the *Euler relation*.[4] It implies that the complex number $\exp(i\eta)$ may be interpreted as a unit vector in the complex plane with a real component $\cos\eta$ and an imaginary component of length $\sin\eta$. Thus, $\exp(i\eta)$ is a unit vector which makes an angle η with the real axis (see Figure 8.1).

Pictures of complex numbers as points or vectors in the complex plane are called *Wessel-Argand diagrams*.[5] Analytic functions can take not only real and imaginary arguments; their arguments can also be complex. Thus, if $\zeta = \xi + i\eta$ is a complex number with ξ, η real,

$$z \equiv e^\zeta = e^\xi e^{i\eta}$$

$$(8.25)$$

is a complex-valued function with real component

$$x = e^\xi \cos\eta$$

$$(8.26)$$

---

[4]Named after the Swiss mathematician Leonhard Euler (1707–1783), who is also credited with introducing the symbol i.

[5]Actually, they are usually simply called *Argand Diagrams* after the Swiss bookkeeper Jean Robert Argand (1768–1822), who made them popular, but they were used somewhat earlier by the Norwegian surveyor Caspar Wessel (1745–1818).

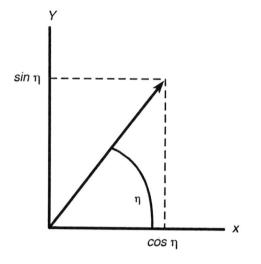

Figure 8.1. The Euler relation $\exp(i\eta) = \cos\eta + i\sin\eta$ means that the complex number $\exp(i\eta)$ can be represented by a unit vector in the complex plane.

and imaginary component

$$y = e^{\xi}\sin\eta. \tag{8.27}$$

One says that the complex function $\exp(\zeta)$ is the *continuation* of the real function $\exp(\xi)$ from the real axis ξ into the complex plane (ξ, η).

Every complex number $z = x + iy$ can be expressed in the form $z = \exp(\zeta) = \exp(\xi)\exp(i\eta)$. The Wessel-Argand diagram of z shows a vector with real and imaginary components $x = \exp(\xi)\cos\eta$ and $y = \exp(\xi)\sin\eta$, respectively, and hence it shows a vector of length $r = \exp(\xi)$ which makes an angle η with the real (x) axis. The length r is the *modulus* or *absolute magnitude* of the complex number z, denoted $|z|$; the angle η is called the *complex phase* of z and is often denoted $\arg(z)$.

The product of two complex numbers

$$z_1 z_2 = (r_1 e^{i\eta_1})(r_2 e^{i\eta_2}) = (r_1 r_2) e^{i(\eta_1 + \eta_2)} \tag{8.28}$$

multiplies the lengths of the corresponding vectors and adds their angles. Thus on a Wessel-Argand diagram, multiplication by a complex number $z = r\exp(i\eta)$ corresponds to a rotation-dilation operation: the original vector is rotated in a counter-clockwise sense by the *phase* η of the complex number, and its length is expanded (dilated) by the factor r. Complex numbers therefore have two interpretations in the complex plane: they are vectors in the plane or they are rotation-dilation operators on such vectors. For example, the unit imaginary i is a unit vector in the y direction, and multiplication by $i = \exp(i\pi/2)$ rotates another vector in the complex plane counter-clockwise by 90°. Since positive and negative real numbers are just complex numbers with phases of 0 and π, multiplication by a positive real number changes only the length and not the angle of the vector corresponding to the complex number multiplied, whereas multiplication by a negative real number not only changes the length of the vector, but also reverses its direction.

The Maple constant I is the imaginary i and `evalc` evaluates complex expressions as the sum of real and imaginary terms. Try

$$
\begin{aligned}
&> \texttt{z1 := 2 + I; z2 := -1 + 2 * I;}\\
&> \texttt{z1 + z2; z1 * z2;}\\
&> \texttt{exp(z1); evalc(exp(z1));}\\
&> \texttt{Re(exp(z1)); Im(exp(z1));}
\end{aligned}
\tag{8.29}
$$

Complex numbers can simplify many manipulations in plane geometry. Thus the real and imaginary parts of the complex product

$$
\begin{aligned}
e^{i\alpha}e^{i\beta} &= e^{i(\alpha+\beta)}\\
&= (\cos\alpha + i\sin\alpha)(\cos\beta + i\sin\beta) = \cos(\alpha+\beta) + i\sin(\alpha+\beta)\\
&= (\cos\alpha\cos\beta - \sin\alpha\sin\beta) + i(\sin\alpha\cos\beta + \cos\alpha\sin\beta)
\end{aligned}
\tag{8.30}
$$

give the useful trigonometric identities for the sine and cosine of the sum of two angles. When $\alpha = \beta$, the result is a special case of *de Moivre's theorem*[6]:

$$
e^{in\alpha} = (\cos\alpha + i\sin\alpha)^n = \cos n\alpha + i\sin n\alpha.
\tag{8.31}
$$

Exercise 8.4. Expand the binomial in the above theorem to derive expressions of $\cos 3\alpha$ and $\sin 3\alpha$ in terms of $\cos\alpha$ and $\sin\alpha$.

---

[6]Named after the French mathematician Abraham de Moivre (1667–1754), who spent most of his life in England.

8.5 *Inverses and the Complex Field

Let $\mathbb{C}$ be the set of complex numbers. The properties established so far are sufficient to make $\mathbb{C}$ a *ring*. For all $u, v, w \in \mathbb{C}$,

(a.1) $u + v \in \mathbb{C}$ ($\mathbb{C}$ is closed under addition);

(a.2) $(u + v) + w = u + (v + w)$ (addition is associative);

(a.3) 0 is an element in $\mathbb{C}$ with the property $u + 0 = 0 + u = u$ (an additive neutral element exists);

(a.4) $-u \in \mathbb{C}$ and obeys $u + (-u) = 0$ (every element has an additive inverse);

(a.5) $u + v = v + u$ (addition is commutative);

(b.1) $uv \in \mathbb{C}$ ($\mathbb{C}$ is closed under multiplication);

(b.2) $(uv)w = u(vw)$ (multiplication is associative);

(c.1) $u(v + w) = uv + uw$ (multiplication is distributive over addition).

The properties (a.1–a.4) define a *group* under addition; (a.5) makes the group *commutative (abelian)*. The two additional conditions

(b.3) $1 \in \mathbb{C}$ is an element such that $1u = u1 = u$ (there exists a neutral multiplicative element), and

(b.4) $uv = vu$ (multiplication is commutative)

make $\mathbb{C}$ a *commutative ring with* (multiplicative) *identity*. However, we have still not defined *division* by complex numbers. Division by an element is equivalent to multiplication by its (multiplicative) inverse, so that defining division is equivalent to finding the inverse of an arbitrary element. Since we know how to divide by any nonzero real number, if given any element $z \in \mathbb{C}$ we can find another element $z^* \in \mathbb{C}$ such that $z\,z^*$ is real and nonzero, then the ratio $z^*/(z\,z^*)$ must be the inverse z^{-1}:

$$z\left(\frac{z^*}{z\,z^*}\right) = \left(\frac{z^*}{z\,z^*}\right) z = 1. \tag{8.32}$$

For $z\,z^*$ to be real, it is necessary and sufficient that its phase be zero (modulo 2π).[7] Therefore, the phase of z^* can be taken to be opposite that of z. If $z = x + iy$ is written $z = r\exp(i\eta)$, then we can define the

---

[7]A number $x + 2n\pi$ where n is any integer is said to be equal to x *modulo* 2π.

complex conjugate of z by[8]

$$z^* = r e^{-i\eta} = x - iy. \tag{8.33}$$

The product $z z^* = z^* z$ is then indeed real: it is the square modulus or length

$$z z^* = r^2 = \left| z^2 \right| = x^2 + y^2. \tag{8.34}$$

Note that the modulus of $z = x + iy$ vanishes if and only if the real and imaginary parts do: $x = y = 0$. Consequently, every nonzero element z of $\mathbb{C}$ has an inverse $z^*/\left| z^2 \right|$, and $\mathbb{C}$ satisfies the additional condition

(b.5) if $z \neq 0$ it has an inverse $z^{-1} \in \mathbb{C}$.

With this additional condition, $\mathbb{C}$ forms a *field*. The rational numbers and the real numbers are also fields.

Try

$$\begin{aligned}
&> \texttt{z := 3 + 4 * I;} \\
&> \texttt{r := sqrt(z * conjugate(z));} \\
&> \texttt{1/z;} \\
&> \texttt{conjugate(z)/r\^{}2;}
\end{aligned} \tag{8.35}$$

In a Wessel-Argand diagram, the complex conjugate z^* of a complex number is found by reflecting the vector representing z in the real axis. Similarly, $-z^*$ gives the reflection of z in the imaginary axis, and $-z$ gives its inversion. In two dimensions, an inversion is equivalent to a $180°$ rotation in the plane.

The projection of z onto the real axis is

$$\Re z = \frac{(z + z^*)}{2}, \tag{8.36}$$

whereas the projection onto the imaginary axis is

$$i \Im z = \frac{(z - z^*)}{2}. \tag{8.37}$$

Note again that what is conventionally called the "imaginary part" of z is the *real* quantity $\Im z = (z - z^*)/(2i)$.

Exercise 8.5. Let z_1 and z_2 be complex numbers. Prove $|z_1 z_2| = |z_1| |z_2|$.

Exercise 8.6. Prove the important relation $(z^{-1})^* = (z^*)^{-1}$.

---

[8]Some authors use the symbol $\bar{z}$ instead of z^*.

8.6   *The Julia and Mandelbrot Sets

Consider the square of a complex number $z_1 = r \exp(i\,\eta_1)$: $z_2 = z_1^2 = r^2 \exp(2\,i\,\eta_1)$. The inverse function to $f : x \to x^2$ is the square-root function $f^{-1} : x \to x^{1/2}$. Thus, z_1 is the square root of z_2. It is not the only square root, however, because $z_2 \exp(2\pi i)$ is the same complex number as z_2, and it has the square root $z_1 \exp(\pi i) = -z_1$. The square root function is said to have two *branches*: the square roots of $z_2 = r_2 \exp(i\eta_2)$ are $r_2^{1/2} \exp[(i\eta_2 + 2\pi i\,m)\,/\,2]$, $m = 0, 1$. Similarly, a complex number $z = r \exp(i\eta)$ generally has n distinct n-th roots, namely

$$z^{1/n} = r^{1/n} \exp[i\,(\eta + 2\pi\,m)\,/\,n],\ 0 \le m < n. \qquad (8.38)$$

Exercise 8.7.   Find the cube roots of $z = 4 + 3i$.

Solution: Rewrite $z = \sqrt{4^2 + 3^2}\,\exp(i\eta) = 5\exp(i\eta)$ where $\tan\eta = 3/4$. The cube roots are then

$$
\begin{aligned}
5^{1/3}\exp(i\Phi_m) &= 5^{1/3}(\cos\Phi_m + i\sin\Phi_m),\\
\Phi_m &= (\eta + 2\pi m)\,/\,3,\ m = 0, 1, 2.
\end{aligned}
\qquad (8.39)
$$

The Maple command

$$> \texttt{solve(rt\^{}3 = 4.0 + 3.0 * I, rt);} \qquad (8.40)$$

finds the roots numerically: $1.67079 + 0.36398i, -1.15061 + 1.26495i$, and $-0.52018 - 1.62894i$.

Consider now what happens to a complex number as it is successively squared. Start with $z_0 = r \exp(i\eta)$, then square z_0 to get z_1, square to obtain z_2, and so on. The general iteration sequence is

$$z_n = z_{n-1}^2 \qquad (8.41)$$

and the first few terms are

$$
\begin{aligned}
z_0 &= r \exp(i\,\eta)\\
z_1 &= r^2 \exp(2\,i\,\eta)\\
z_2 &= r^4 \exp(4\,i\,\eta)\\
z_3 &= r^8 \exp(8\,i\,\eta)\,,
\end{aligned}
\qquad (8.42)
$$

and after n iterations the result is

$$z_n = r^{2^n} \exp(2^n i \eta). \tag{8.43}$$

If $r < 1$, successive iterations will spiral into the origin, whereas if $r > 1$, the progression will wander off to infinity. With $r = 1$, the points will jump around on the circle of radius 1 in the complex plane. The set of points z_0 for which the sequence does *not* diverge is called the *Julia set for the point* $C = 0$. It is the closed set $\{z \ni |z| \leq 1\}$, that is, the unit disk centered at the origin. That may not be too exciting, but Julia sets become more interesting for other points C.

There is a Julia set for every point C in the complex plane. It is the set

$$\{z_0 \ni \lim_{n \to \infty} |z_n| \leq M\} \tag{8.44}$$

where M is a real finite constant ($M = 2$ is typical) and the iteration scheme is

$$z_n = z_{n-1}^2 + C.$$

For any point C the there will be solutions with $z_n = z_{n-1}$, namely solutions $z_n = 1/2 \pm i\sqrt{C - 1/4}$, to the quadratic equation $z^2 - z + C = 0$. But there will also be solutions, for example, in which $z_n = z_{n-2}$, others for which $z_n = z_{n-3}$, and so on. For points C close to the origin, the Julia sets tend to be connected circular regions, but for distant values of C, the sets comprise disconnected points. The sets become most interesting for intermediate values of C.

Benoit Mandelbrot (1924–) asked: for what values of C is the origin in the Julia set? The set of such C values is now known as the *Mandelbrot set*. It turns out to be a finite region of the complex plane with an infinitely long boundary known as a *fractal*. It displays infinite variety at all scales of magnification; in fact, the boundary looks roughly the same no matter how much it is magnified. This is a characteristic of fractals known as *self-similarity*. It is instructive and somewhat awe-inspiring to see how much intricate beauty can arise from extremely simple algorithms with complex numbers.

Many other fractals have been generated from different iteration schemes. A good selection of these may be generated and viewed at different magnifications in the computer program Fractint.

8.7 *Fun with "Fractint"

Fractint is a free package of computer programs to generate color graphics images of fractals and related designs. It was written by four main authors (Bert Tyler, Tim Wagner, Mark Peterson, and Pieter Branderhorst) and a couple dozen contributors: hackers, mathematicians, computer scientists, and others in what they named the "Stone Soup Group," who all communicate by electronic mail. By using 16-bit and 32-bit integer arithmetic efficiently, Fractint produces colorful images at an amazing tempo and is a stunning example of what can be produced by computer-age cooperation.

The programs reside in a half-Mbyte compressed file which can be downloaded from CompuServe and elsewhere, as described in the Preface. It is suggested that you copy the compressed file to a directory fractal on your hard disk, and then expand it there. If you have the self-expanding DOS file FRAINT.EXE, you can simply execute it. It will automatically expand into a collection of files which occupy around 2 Mbytes. Then from the DOS prompt, you can change to the fractal directory and type demo to see a sample. Run Fractint for more controlled playing. The IBM VGA driver works well, and although it doesn't exploit the full resolution of SVGA computers, it is fast.

Be sure to investigate the standard Mandelbrot set. Use $\boxed{\text{Pg Up}}$ to reduce the frame and the cursor keys to center it on an interesting feature; then hit $\boxed{\leftarrow\text{Enter}}$ to zoom in for an expanded view of what's inside the frame. Use $\boxed{\text{Ctrl}}$ $\boxed{\leftarrow\text{Enter}}$ to zoom out. You can repeat this procedure many times to get magnification factors of over a trillion. Return to the main menu by striking the $\boxed{\text{Esc}}$ key; hit the key again to exit. Extensive on-line help is available through the F1 function key, and 140 pages of documentation can also be downloaded for printout. (See especially Section 8.1: A Little History, and Appendix E: Bibliography.)

You can make and distribute copies of the program, but you are not permitted to sell them. The copyright is retained by the program's authors.

8.8 *Uniform Circular Motion and Oscillation

Uniform circular motion in the xy plane can be described by a complex variable with time dependence

$$z(t) = z(0)e^{-i\omega t} \tag{8.45}$$

where $z(0) = r\exp(i\eta)$ is the initial position; that is, r is the radius of the motion and η is the initial angle of the position vector with respect to the x-axis. Recall (Section 8.4) that the factor $\exp(-i\omega t)$ rotates $z(0)$ clockwise from the x-axis by an angle ωt which is increasing linearly in time. Thus, ω is the *angular velocity* of the motion in a clockwise direction. The real and imaginary parts of $z(t)$, proportional to $z \pm z^*$, respectively, are the x and y components of the position vector as a function of time. Since the time-derivative is real and linear,

$$\frac{d}{dt}(z \pm z^*) = \frac{dz}{dt} \pm \left(\frac{dz}{dt}\right)^*, \tag{8.46}$$

and the velocity components dx/dt, dy/dt are the real and imaginary parts of

$$\dot{z}(t) \equiv \frac{dz}{dt} = -i\omega z(t) = \omega z(0)e^{-i(\omega t + \pi/2)}, \tag{8.47}$$

namely

$$\dot{x}(t) = \omega y(t), \quad \dot{y}(t) = -\omega x(t). \tag{8.48}$$

However, the complex form (8.47) may be more telling: one sees that the velocity is always 90° ahead of the position.

Complex notation is also convenient for modeling simple-harmonic motion, and orbits in the complex plane are closely related to those in phase space. A second derivative of $z(t)$ (8.45) with respect to time gives the acceleration:

$$\ddot{z}(t) \equiv \frac{d^2 z}{dt^2} = -\omega^2 z(t) \tag{8.49}$$

which is seen to be directed opposite to the position vector, and hence inward toward the origin. Taking the real part of the above equation of motion, we obtain the equation

$$\ddot{x}(t) = -\omega^2 x(t) \qquad (8.50)$$

for the motion of a simple harmonic oscillator. For systems with a linear restoring force $F = m\ddot{x} = -kx$ and mass m, the angular frequency can be associated with $\sqrt{k/m}$. Thus the motion of the simple harmonic oscillator must be given by

$$x(t) = \Re\{z(0)\exp(-i\omega t)\}. \qquad (8.51)$$

The two constants in $z(0)$ (either $x(0)$, $y(0)$ or r, η) are determined by initial conditions for $x(0)$ and $\dot{x}(0) = \omega\, y(0)$. Since the velocity is $\dot{x}(t) = \omega\, y(t)$ [see equation (8.48)], the phase-space trajectory of $\dot{x}(t)$ vs. $x(t)$ is, within the scaling factor ω of the imaginary axis, identical to the plot of $z(t)$ in the complex plane.

The fact that the solution (8.51) is a general solution of the second-order ordinary differential equation is indicated by the two independent integration constants to be determined by initial conditions. It can also be shown with a direct solution of the differential equation for the oscillator

$$(D^2 + \omega^2)\, x(t) = 0 \qquad (8.52)$$

by factoring the differential operator into first-order terms with imaginary roots

$$(D - i\omega)(D + i\omega)\, x(t) = 0 . \qquad (8.53)$$

Since the two factors commute, the solution can be expressed as a linear combination of the solutions to $(D - i\omega)\, x(t) = 0$ and $(D + i\omega)\, x(t) = 0$:

$$x(t) = c_1 e^{i\omega t} + c_2 e^{-i\omega t} . \qquad (8.54)$$

By constraining $x(t)$ to be real, we obtain the solution (8.51) above. Maple's solutions were found in the last chapter:

$$> \texttt{dsolve}((\texttt{D@@2})(\texttt{x})(\texttt{t}) + \texttt{omega}\hat{}\,2 * \texttt{x}(\texttt{t}) = 0, \texttt{x}(\texttt{t})); \qquad (8.55)$$

The initial conditions again determine the coefficients; for physical systems, they should also guarantee that the solution is real.

Damping proportional and opposite to the velocity is added by including a first-order term $\Gamma \dot{x}(t)$:

$$(D^2 + \Gamma D + \omega^2)x(t) = 0. \qquad (8.56)$$

We can again factor the differential operator into commuting, first-order operators:

$$(D + \frac{\Gamma}{2} + i\omega')(D + \frac{\Gamma}{2} - i\omega')x(t) = 0 \qquad (8.57)$$

where $\omega' = \sqrt{\omega^2 - \Gamma^2/4}$. The roots are now complex. The solution must be a real linear combination of

$$\exp(-\Gamma t/2)\exp(\pm i\,\omega' t). \qquad (8.58)$$

Alternatively, for subcritical damping ($\Gamma < 2\omega$) the solution can be written as the real part

$$x(t) = \Re\{z(0)\exp(-\Gamma t/2 - i\omega' t)\} \qquad (8.59)$$

of $z(t) = z(0)\exp(-\Gamma t/2 - i\omega' t)$, where the real and imaginary parts of $z(0)$ are determined by the initial conditions.

This is equivalent to the solutions found in Section 7.6 with Maple. If no initial conditions are specified, Maple gives a general solution with two integration constants:

```
> ODE := (D@@2)(x)(t) + Gamma * D(x)(t) + omega^2 * x(t) = 0;
> dsolve(ODE, x(t));
```
$$(8.60)$$

When the damping Γ is small compared to the natural oscillator frequency ω, the trajectory in phase space, like that in the complex plane, spirals in a clockwise sense into the *attractor* at the origin. However, the velocity $\dot{x}$ is no longer ωy, but rather $\omega \hat{v} \cdot \mathbf{r}$ where

$$\omega \hat{v} = -\frac{\Gamma}{2}\hat{x} + \omega'\hat{y} \quad , \qquad \Gamma < 2\omega , \qquad (8.61)$$

which is at an angle $\alpha = \arcsin \frac{\Gamma}{2\omega}$ clockwise from the y-axis. When the damping is critical, $\Gamma = 2\omega$ and $\omega' = 0$. Then, as at supercritical values of the damping constant, $\Gamma > 2\omega$, there is no spiraling: $z(t)$ moves radially and eventually shrinks exponentially to zero.

8.9   *The Pendulum: A Nonlinear System

The pendulum is a nonlinear oscillator (see Section 3.2) with an equation of motion

$$\ddot{\theta} + \Gamma\dot{\theta} + \omega^2 \sin\theta = 0 \tag{8.62}$$

where $\omega = \sqrt{g/l}$ is the frequency of the pendulum in rad/s [and hence 2π times the frequency in cycles/s or Hz] for small-amplitude oscillations, θ is the angle the pendulum makes with the gravitational acceleration $\mathbf{g}$, and Γ is a damping term which has been added as above. Since the length l of the pendulum is fixed, we may think of the system as a small mass on a light, rigid rod which is free to rotate about one end in a vertical plane. Let's first investigate the phase-space trajectories when the damping is negligible; the system is then *conservative*: its total energy is a constant. Phase space, introduced in Section 7.6, has axes of $\dot{\theta} \equiv d\theta/dt$ and θ. We can find a relation between $\dot{\theta}$ and θ by integrating the force equation over θ :

$$\int d\theta\ddot{\theta} = \int \dot{\theta}\, d(\dot{\theta}) = -\omega^2 \int d\theta \sin\theta . \tag{8.63}$$

We find

$$\omega^{-2}\dot{\theta}^2 - 2\cos\theta = 2\epsilon , \tag{8.64}$$

where the integration constant $\epsilon = E/(mgl)$ is the total energy in units of mgl :[9]

$$\epsilon = \frac{E}{mgl} = \frac{\frac{1}{2}ml^2\dot{\theta}^2 - mgl\cos\theta}{mgl} = \frac{\dot{\theta}^2}{2\omega^2} - \cos\theta. \tag{8.65}$$

The minimum value of ϵ is -1: it gives a motionless pendulum hanging straight down in equilibrium. For values $-1 < \epsilon < 1$ the energy determines the maximum angle reached by the swinging pendulum. For

---

[9]It is generally efficient when plotting or computing numerical values to eliminate as many paramters as possible by appropriate choices of dimensionaless variables. Thus we choose $\dot{\theta}$ in units of the natural frequency ω, or equivalently time in units of ω^{-1}, and energy in units of mgl.

$\epsilon > 1$, $\dot{\theta}$ is never zero and the pendulum has enough energy to swing over the top.

The phase-space trajectories are given by

$$\dot{\theta}/\omega = \pm\sqrt{2(\epsilon + \cos\theta)}. \tag{8.66}$$

Plot some of them with the commands

```
> sr := epsilon- > sqrt(max(0, 2 * (epsilon + cos(theta))));
> sq := sr(2), -sr(2), sr(1), -sr(1), sr(0), -sr(0), sr(-.7), -sr(-.7)
> plot({sq}, theta = -Pi..2 * Pi);
```
$$\tag{8.67}$$

A collection of phase-space trajectories such as given by the above commands (see Figure 8.2) shows the motion for any initial condition $\left(\theta, \dot{\theta}\right)_{t_0}$ in a system described by a second-order differential equation. It is called a *phase portrait* of the system.

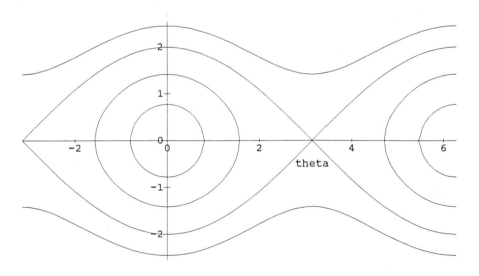

Figure 8.2. Phase portrait of undamped pendulum. The vertical axis gives the angular velocity $\dot{\theta}$ in units of the natural frequency ω.

To see which trajectory corresponds to which value of the energy, consider the value of the angular velocity $\dot{\theta}$ when $\theta = 0$. Note that the trajectory for $\epsilon = 1$ separates trajectories of qualitatively different types: the ones at smaller ϵ are all bound orbits about the attractor (a point of *stable equilibrium*) at the origin, whereas the ones with larger ϵ go over the top and never return to the same θ value. A microscopic change in the initial condition can cause a large qualitative change in the orbit. This is characteristic of the *chaos* often associated with nonlinear systems. The $\epsilon = 1$ trajectory which separates the bounded paths from the free ones is called a *separatrix*. Only the separatrices pass through the fixed points at $\dot{\theta} = 0$, $\theta = \pm\pi$. These are points of *unstable equilibrium*: any small displacement of the system from such a point results in a force which drives the system still farther away. They are also known as *saddle points* because of the way the trajectories enter along one direction and exit along another. Note that no phase-space trajectory passes smoothly through these points: sharp turns are invariably required by the fact that phase-space trajectories with $\dot{\theta} > 0$ ("above the θ axis") move to the right, whereas those with $\dot{\theta} < 0$ ("below the θ axis") move to the left.

When damping is present, the system always loses energy and all the trajectories eventually end at an attractor. There is no analytic solution for the nonlinear damped pendulum, but numerical solutions are possible. If the damping constant is small, the system should basically follow the conservative trajectories but with an ever decreasing value of ϵ. Near the stable fixed points, the force is nearly linear in the displacement, and the trajectories are like those found for the linear damped oscillator (Section 8.8). The case of critical damping is treated in problem 7.

Maple has a special command for plotting phase portraits of systems which may need to be computed numerically. The procedure `phaseportrait` is in the `DEtools` package (see also Problem 8 of Chapter 7). To plot the phase portrait of the damped pendulum (8.62) with

$\dot{\theta}$ and Γ in units of ω, send Maple the instructions

```
> with(DEtools) :
> Gamma := 0.1; ics := {[0, 0, 1], [0, 0, 2], [0, 0, 2.5], [0, 0, 3]};
> deq := [y, −Gamma * y − sin(x)];
> phaseportrait(deq, [x, y], 0..18, ics, stepsize = .2);
```
(8.68)

We have used x for θ and y for $\dot{\theta}$. The first argument gives the differential equation in the form of a list of the functional expressions for

$$[dx/dt, dy/dt] .$$
(8.69)

The second argument, $[x, y]$, is a list of the variables to be plotted; the third is the domain of t values for which the trajectories are to be drawn; and the fourth is the set of initial conditions, each given as a list of the initial values of t, x, y. Starting with values given in each of the initial conditions, the command numerically integrates and plots the differential equations over the domain specified. An optional parameter has been added to control the stepsize.[10]

8.10 *Problems

1. Find the real and imaginary parts of the functions

$$\begin{aligned} \sinh z &= \sinh(x + iy) \\ \cosh z &= \cosh(x + iy). \end{aligned}$$
(8.70)

(a) Prove that the complex conjugate of a product of two complex numbers is the product of their complex conjugates, $(z_1 z_2)^* = z_1^* z_2^*$, and that the complex conjugate of a sum is the sum of the complex conjugates, $(z_1 + z_2)^* = z_1^* + z_2^*$.

(b) Let $f(x)$ be a real analytic function of the real variable x. Prove that its continuation to the complex plane obeys

$$f^*(z) = f(z^*).$$
(8.71)

---

[10]For more information, ask Maple for help:
```
> ?phaseportrait
```

2. Let $z_1 = x_1 + iy_1$ and $z_2 = x_2 + iy_2$ be complex numbers.

   (a) Show that the product $z_1 z_2^*$ is invariant under rotations, that is, under any operation which rotates z_1 and z_2 by the same angle in the complex plane.

   (b) Relate the real and imaginary parts of the product $z_1 z_2^*$ to the dot and cross products of the two-dimensional vectors $\mathbf{r}_1 = (x_1, y_1)$, $\mathbf{r}_2 = (x_2, y_2)$. In particular, prove that

   $$z_1 z_2^* = \mathbf{r}_1 \cdot \mathbf{r}_2 - i \mathbf{r}_1 \times \mathbf{r}_2 \qquad (8.72)$$

   where in a two-dimensional vector space, the cross product is the scalar

   $$\mathbf{r}_1 \times \mathbf{r}_2 := x_1 y_2 - y_1 x_2, \qquad \mathbf{r}_1, \mathbf{r}_2 \in \mathbb{R}^2 . \qquad (8.73)$$

3. Evaluate the following, expressing each in the form $x + iy$:

   (a) The four fourth roots of 1
   (b) $\sqrt{-i}$
   (c) $(3 + I * 4)/(1 + i)$

4. Consider the complex form of the general solution to the simple harmonic oscillator, namely

   $$z(t) = c_1 \exp(i\omega t) + c_2 \exp(-i\omega t) .$$

   (a) Re-express the right-hand side of this relation in terms of the trigonometric functions $\sin$ and $\cos$. Show that the relation can be written in the form

   $$z(t) = a \cos(\omega t) + ib \sin(\omega t) ,$$

   and find a, b in terms of c_1, c_2. Let $z(t) = x(t) + iy(t)$, assume the coefficients c_1, c_2 are real and positive, and solve for $x(t)$ and $y(t)$.

(b) Eliminate ωt by summing $\cos^2(\omega t) + \sin^2(\omega t) = 1$ and show that the resulting equation for the position (x, y) describes an ellipse in the complex plane:

$$(x/a)^2 + (y/b)^2 = 1 .$$

(c) How should $z(t)$ as given above be modified in order to represent an ellipse whose semimajor axis makes an angle ϕ with respect to the x-axis?

5. Consider two identical simple-harmonic oscillators of mass m and displacements z_1 and z_2 which vibrate with angular frequency ω. Now couple them together with a weak spring of force constant $\frac{1}{2}m\Omega^2$ where $\Omega \ll \omega$ so that they obey the coupled equations of motion

$$(D^2 + \omega^2) z_1 + \tfrac{1}{2}\Omega^2 (z_1 - z_2) = 0$$
$$(8.74)$$
$$(D^2 + \omega^2) z_2 + \tfrac{1}{2}\Omega^2 (z_2 - z_1) = 0$$

(a) Show by substitution that there exist two solutions of the form $\{z_1(t) = z_1(0)\exp(-i\nu t), \ z_2(t) = z_2(0)\exp(-i\nu t)\}$. These solutions are called *natural modes of vibration* or *eigensolutions*, and they describe a time-dependence in which the whole system vibrates at a single frequency. Demonstrate that one of the solutions has angular *eigenfrequency* $\nu = \pm\omega$ and relative amplitudes $z_1 = z_2$.

(b) Find the eigenfrequency and relative amplitudes for the other natural mode.

(c) The system is said to have two *degrees of freedom*, represented by the variables $\{z_1, z_2\}$. One can show that any solution to (8.74) can be written as a linear combination of the eigensolutions. Determine the linear combination corresponding to the initial conditions $\{z_1(0) = 1, z_2(0) = \dot{z}_1(0) = \dot{z}_2(0) = 0\}$ and plot the resultant motion. [Hint: You must either take linear combinations with both positive and negative frequencies, or you should remember to take the real parts, to ensure that the physical solutions are real.]

6. Use Maple to determine whether or not the point $C = 0.7 + 0.8i$ belongs to the Mandelbrot set. Show your command sequence. [Hint: 5 iterations is enough to tell.]

7. Let Maple find solutions of the damped linear harmonic oscillator (Section 8.8) for the case of *critical damping*: $\Gamma = 2\omega$. Verify by hand that Maple's solutions do satisfy the second-order differential equation, and then plot phase-space trajectories of $\dot{\theta}/\omega$ vs. θ for initial values $(\theta, \dot{\theta}/\omega) = (0, 1)$, $(0, 4)$, and $(0, 9)$. [You can use Maple's parametric plots for this purpose, in particular plots of the form

$$> \texttt{plot([theta(t), D(theta)(t), t = 0..10]);]} \qquad (8.75)$$

You may also use the `phaseportrait` command. Produce either a printout of the plots or a fair sketch of them for the three trajectories. Finally, find the conditions on the initial values for which the phase-space trajectory is a straight line.

8. Consider a pair of vectors z and c in the complex plane.

   (a) Describe the relative length and orientation of the vector cz^*c^{-1*}. [Hint: use the polar form $r \exp(i\phi)$ for the complex numbers.]

   (b) Show that the part of z parallel to c is $\frac{1}{2}(z + cz^*c^{-1*})$. Verify that this relation gives the real and imaginary portions of z when $c = 1$ and $c = i$, respectively.

   (c) Show that the part of z perpendicular to c is $\frac{1}{2}(z - cz^*c^{-1*})$.

8.11   Chapter Summary

8.11.1   Concepts

imaginary numbers

complex numbers, complex variables

modulus, phase

the complex plane

real and imaginary components

real and imaginary axes

group, ring, field

vector spaces

Euler relation

de Moivre's theorem

Wessel-Argand diagrams

complex powers and roots

Julia and Mandelbrot sets

critical damping

conservative systems

phase-space trajectories

phase portraits

separatrix

stable and unstable equilibrium

saddle points

8.11.2 Maple Commands

```
I
factor
while...do...od;
evalc
```

Re, Im

conjugate

abs

with(DEtools)

phaseportrait

Chapter 9

Vector Algebra of Physical Space

In this chapter, we study the vector algebra of 3-dimensional space. The term "algebra" is meant here in its mathematical sense, so that in addition to the usual vector-space manipulations, an associative multiplication of vectors is required. Relatively simple considerations lead us to what is called the *geometric algebra* (or *Clifford algebra*) of 3-dimensional space, also known as the Pauli algebra. The standard matrix representation of this algebra replaces basis vectors by Pauli spin matrices (and hence the name "Pauli algebra"), but specific representations encumber the mathematics with unnecessary baggage; it is usually simpler to work directly in the algebra in component-free notation without reference to any matrices.

A general element of the algebra is called a *cliffor* and comprises both scalar and 3-dimensional vector parts. It can be viewed mathematically as a vector in 4-dimensional space. Cliffors commute under multiplication only if their spatial-vector parts are parallel. The field of complex numbers arises naturally as the center of the algebra, that is, as the part which commutes with all cliffors. Real cliffors in the Pauli algebra are called *paravectors*. They form a 4-dimensional *paravector space* which provides a framework for the description of relativistic phenomena. A study of invertible elements of the algebra leads to a scalar product which is identical to the special vector norm used in relativity (the Minkowski spacetime norm). *Lorentz transformations* (or *spacetime rotations*) are linear transformations which leave the norm invariant. Through the use of cliffors, the Pauli algebra is able to provide a covariant description of relativistic physics, while the split of real cliffors into scalar and vector parts by an observer maintains the qualitative difference between time and space variables in any given frame.

Section 9.1 introduces the Pauli algebra $\mathcal{P}$, and Section 9.2 relates the algebraic product of vectors to the usual dot and cross products and shows how the trivector element of the algebra plays the role of i. The geometric interpretation of algebraic products is discussed in the following section, and the algebraic representation of rotations and reflections is emphasized with examples in Section 9.4. Some aspects of the algebraic structure, in particular the subalgebras of the Pauli algebra, are considered in Section 9.5, before we study applications in special relativity and electromagnetic theory in sections 9.6 and 9.7. Maple procedures for the Pauli algebra are discussed in Section 9.8.

9.1   Vector Algebra

The description of motion in 3-dimensional (Euclidean) space $\mathbb{R}^3$ requires many quantities which are usually taken to be vectors; relative positions, velocities, accelerations, forces, electric and magnetic fields, and others are part and parcel of the familiar merchandise of physics. Every vector of $\mathbb{R}^3$ can be written as a linear combination

$$\mathbf{a} = a^j \mathbf{e}_j \tag{9.1}$$

(the convention of summing over repeated indices is adopted here) of the orthonormal basis vectors $\mathbf{e}_j$, $j = 1, 2, 3$:

$$\mathbf{e}_j \cdot \mathbf{e}_k = \delta_{jk} \ . \tag{9.2}$$

Traditional vector notation in physics can be traced back to J. Willard Gibbs (1839–1903) and Oliver Heaviside (1850–1925)[1]; it relies on the rather restricted manipulations of vectors required in a vector space with a Euclidean inner product: vector addition $\mathbf{a} + \mathbf{b}$; the dot,

---

[1]P. J. Nahin, "Oliver Heaviside," *Sci. Am.* **262** (6), 122–129 (1990).

scalar, or inner product of vectors $\mathbf{a} \cdot \mathbf{b} = a^j b^k \mathbf{e}_j \cdot \mathbf{e}_k = |\mathbf{a}|\,|\mathbf{b}|\cos\theta$, where θ is the angle between $\mathbf{a}$ and $\mathbf{b}$; and multiplication $\lambda\mathbf{a}$ of vectors $\mathbf{a}$ by scalars λ. One additional product has been added to the standard repertoire of vector calculations in three dimensions, namely the cross product

$$\mathbf{a} \times \mathbf{b} = a^j b^k \epsilon_{jkl} \mathbf{e}_l \tag{9.3}$$

where ϵ_{jkl}, called the Levi-Civita symbol, is $+1$ (-1) if j, k, l is a cyclic (anticyclic) permutation of 1,2,3 and vanishes if any two of the indices are equal (see Chapter 4). Products of vectors are traditionally represented in terms of Cartesian components. However, the use of components requires the specification of an axis system, at least implicitly, and defeats some of the beauty of the component-free vectors introduced by Gibbs.

The operations defined among vectors in a vector space are quite limited compared to those available for scalars. William Rowan Hamilton (1805–1865), in his development of quaternions[2], sought an algebra which would give to vector computations in three-dimensional space much of the computational power one has in a two-dimensional vector space with complex numbers (see Chapter 8). We follow here the somewhat more general approach of William Kingdon Clifford (1845–1879), who was able to unite the efforts of Hamilton and the German mathematician Herman Grassmann (1809–1877). Much of the credit for recognizing and emphasizing the importance of geometric algebras in physics in more recent times must be accorded David Hestenes.[3]

---

[2]Quaternions are defined at the end of Section 9.5. They are the topic of Hamilton's two volumes *Elements of Quaternions*, 1866 3rd edition, reprinted by C. J. Jolly, ed. (Chelsea Publ. Co., New York, 1969).

[3]See, for example, D. Hestenes, *Space-Time Algebra* (Gordon and Breach, New York, 1966); D. Hestenes, "Vectors, spinors, and complex numbers in classical and quantum physics," *Am. J. Phys.* **39**, 1013–1027 (1971); D. Hestenes, "Proper particle mechanics," *J. Math. Phys.* **15**, 1768–1777 (1974); D. Hestenes, "Proper dynamics of a rigid point particle," *J. Math. Phys.* **15**, 1778–1786 (1974); D. Hestenes and G. Sobczyk, *Clifford Algebra to Geometric Calculus* (Reidel, Dordrecht, 1984); and D. Hestenes, *New Foundations for Classical Mechanics* (Reidel, Dordrecht, 1987). See also J. S. R. Chisholm and A. K. Common, editors, *Clifford Algebras and Their Applications in Mathematical Physics I* (Reidel, Dordrecht 1986); A. Micali, R. Boudet, and J. Helmstetter, editors, *Clifford Algebras and Their Applications in Mathematical Physics II* (Reidel, Dordrecht 1992); F. Brackx, R. Delanghe, and H.

The essential ingredient missing from the vector calculations of Gibbs and Heaviside is an associative, invertible product of vectors which is distributive over vector addition. Neither the dot product nor the cross product by itself is invertible, since knowing both **a** and **a** · **b** (or both **a** and **a** × **b**) does not allow one to determine **b**. Further, neither product is associative. Let's postulate an associative, invertible product of vectors **a**, **b**, which for simplicity we indicate by **ab**, and try to work out its properties. We do not assume the product to be commutative: **ab** may not equal **ba**, but we do assume it to be distributive over addition: **a** (**b** + **c**) = **ab** + **ac** and (**a** + **b**) **c** = **ac** + **bc**; and to commute with scalar multiplication: **a** (λ**b**) = λ (**ab**) = (λ**a**) **b**, where λ is any real scalar. It is sufficient to find the relations for products of the three unit basis vectors, since those for linear combinations thereof will then follow easily. Later, we can make the notation component-free.

It is useful to anticipate some results. In three-dimensional space, we can divide elements into four geometric classes: scalars (elements of zero dimensions), lines (one-dimensional elements), surfaces (two-dimensional elements), and volumes (three-dimensional elements). Directed lines are, of course, just vectors. Algebraic products of two vectors contain information about the plane in which the vectors lie, the area determined by them, and the direction of rotation or circulation from one vector to the other. Products of three vectors contain information about the volume, say, of the parallelepiped whose edges are given by the vectors and the handedness implied by the ordering of the vectors.

9.2 Algebraic Products of Vectors

To determine the basic rule for vector products, it suffices to consider the product of a vector with itself. The square of a real vector cannot contain a vector part, because the only direction singled out is that of

Serras, editors, *Clifford Algebras and Their Applications in Mathematical Physics III* (Reidel, Dordrecht 1993); B. Jancewicz, *Multivectors and Clifford Algebras in Electrodynamics* (World Scientific, Singapore 1989); and Z. Oziewicz, B. Jancewicz, and A. Borowiec, editors, *Spinors, Twistors, Clifford Algebras and Quantum Deformations* (Reidel, Dordrecht, 1993).

a of itself, but $\mathbf{a}^2 = (-\mathbf{a})^2$ so that there is no reason to choose **a** over $-\mathbf{a}$. Furthermore, no volume and no plane containing **a** is uniquely specified by $\mathbf{a}^2$. We are led to choose $\mathbf{a}^2$ to be a scalar, and the obvious choice is

$$\mathbf{a}^2 = a^j a^k \mathbf{e}_j \mathbf{e}_k = \mathbf{a} \cdot \mathbf{a} = a^j a^k \mathbf{e}_j \cdot \mathbf{e}_k. \tag{9.4}$$

This constraint gives the basic multiplication rule for the vector algebra. Since it holds for any vector components $a^j a^k$, we can equate the factors multiplying $a^j a^k$ to obtain

$$\begin{aligned} \mathbf{e}_j^2 &= 1 \\ \mathbf{e}_j \mathbf{e}_k + \mathbf{e}_k \mathbf{e}_j &= 0 \end{aligned} \tag{9.5}$$

which, with the help of the Kronecker delta (see Chapter 4), can be summarized in the single equation:

$$\mathbf{e}_j \mathbf{e}_k + \mathbf{e}_k \mathbf{e}_j = 2\delta_{jk}. \tag{9.6}$$

The product of any number of vectors is completely determined by the rule (9.6). The product of two orthogonal vectors is called a *bivector* and, as discussed below, gives information about the plane in which the vectors lie and the direction of the rotation in that plane which takes the first vector into the direction of the second. There are just three linearly independent bivectors which can be formed from products of the $\mathbf{e}_j$: $\mathbf{e}_1\mathbf{e}_2, \mathbf{e}_2\mathbf{e}_3$, and $\mathbf{e}_3\mathbf{e}_1$. In the product of three orthonormal basis vectors, if any two of the vectors are the same, the result is (to within a sign) the remaining one:

$$\mathbf{e}_1\mathbf{e}_1\mathbf{e}_2 = \mathbf{e}_2 = -\mathbf{e}_1\mathbf{e}_2\mathbf{e}_1 \tag{9.7}$$

and so on. In three dimensions, the only triple products of basis vectors which are not linear combinations of the basis vectors themselves are permutations of the trivector $\eta := \mathbf{e}_1\mathbf{e}_2\mathbf{e}_3$, which may be interpreted as the volume of the unit cube of sides $\mathbf{e}_1, \mathbf{e}_2$, and $\mathbf{e}_3$ with a handedness given by the ordering $\mathbf{e}_1, \mathbf{e}_2$, and $\mathbf{e}_3$. That's it! That exhausts the possible products of basis vectors. Any product of four basis vectors of 3-dimensional space must contain at least two identical factors and can, with the help of (9.6), be re-expressed as a product of two basis vectors.

Similarly, all higher-order products of the e_j can be reduced to linear combinations of the 8 basis elements $\{1, e_1, e_2, e_3, e_1e_2, e_2e_3, e_3e_1, \eta\}$.

Note from the rule (9.6) that every real unit vector is its own inverse $e_j^{-1} = e_j$. Furthermore, the three distinct bivector basis elements can be written as products of the trivector η with a vector, for example,

$$e_1e_2 = e_1e_2e_3e_3^{-1} = \eta e_3^{-1} = \eta e_3 = \eta\epsilon_{12k}e_k.$$

More generally,

$$e_je_k - e_ke_j = 2\eta\epsilon_{jkl}e_l. \tag{9.8}$$

The vector $\epsilon_{jkl}e_l$ is called the (Hodge) dual of the bivector e_je_k, $j \neq k$, and is normal to the surface represented by the bivector.[4] Although the vectors and bivectors do not generally commute with each other, one can verify that the trivector η does commute with all the basis forms in $\mathcal{P}$, for example, by (9.6)

$$e_1\eta = e_1e_1e_2e_3 = e_1e_2e_3e_1 = \eta e_1. \tag{9.9}$$

Similarly, η commutes with any vector and hence with any product of vectors. Therefore, η commutes with every cliffor in $\mathcal{P}$. Furthermore,

$$\eta^2 = e_1e_2e_3e_1e_2e_3 = e_1e_2e_1e_2 = -1. \tag{9.10}$$

In the vector algebra of 3-dimensional space, the trivector η has the same mathematical properties as the unit imaginary i, and we can therefore use the symbols interchangeably[5]: $\eta = i$. This is how the imaginary i arises in the Pauli algebra.[6]

---

[4]Only in three dimensions can the orientation of a plane be specified by a vector. In four dimensions, for example, the directions normal to a plane define another, orthogonal plane, and the dual of a bivector is then another bivector.

[5]Certainly another choice is $\eta = -i$. The change from $\eta = i$ to $\eta = -i$ is equivalent to the transformation from a right-handed system to a left-handed one.

[6]More generally, such vector algebras, called *Clifford algebras*, can be found in spaces of n dimensions and various signatures. The volume basis element η is then the product of all n orthonormal basis vectors: $\eta = e_1e_2\cdots e_n$. It will commute with vectors if n is odd and anticommute if n is even. Suppose p of the basis vectors square to $+1$ and $q = n - p$ of them square to -1. It can be shown that the volume element η commutes with all elements and squares to -1 if and only if $p - q = 3 \mod 4$, in other words iff $(p, q) = (0, 1), (1, 2), (2, 3), (3, 0), (4, 1)$, etc.

Exercise 9.1. Use vector algebra to evaluate

(a) $(\mathbf{e}_1 + \mathbf{e}_2)^2$

(b) $(\mathbf{e}_1 + \mathbf{e}_2)(\mathbf{e}_1 - \mathbf{e}_2)$

(c) $\mathbf{e}_1\mathbf{e}_2\mathbf{e}_3\mathbf{e}_2\mathbf{e}_1$

Answer: $2, -2i\mathbf{e}_3, \mathbf{e}_3$.

From products of 3-dimensional vectors, we have thus generated an 8-dimensional real vector space whose elements are real linear combinations of eight basis elements in four subspaces:

- one scalar $(1 \in \mathcal{V}_0)$,

- three basis vectors $(\mathbf{e}_j \in \mathcal{V}_1)$,

- three bivectors $(i\mathbf{e}_k \in \mathcal{V}_2)$, and

- the trivector $(i \in \mathcal{V}_3)$.

Since we can identify $\mathcal{V}_3 = i\mathcal{V}_0$ and $\mathcal{V}_2 = i\mathcal{V}_1$, the 8-dimensional real space V comprising real linear combinations of elements of all four subspaces, which we can indicate by $V = \mathcal{V}_0 \oplus \mathcal{V}_1 \oplus \mathcal{V}_2 \oplus \mathcal{V}_3$, can also be considered a 4-dimensional complex vector space. Every element is then a complex linear combination of the four basis elements $\{1, \mathbf{e}_1, \mathbf{e}_2, \mathbf{e}_3\}$ of the *paravector space* $\mathcal{V}_0 \oplus \mathcal{V}_1$. The algebra of real three-dimensional vectors thus generates a complex four-dimensional space and provides a natural introduction of complex numbers into the real world of physics.[7] By adding (9.6) and (9.8) we can also express the product rule by

$$\mathbf{e}_j\mathbf{e}_k = \delta_{jk} + i\epsilon_{jkl}\mathbf{e}_l. \tag{9.11}$$

This product rule is a relation also obeyed by the Pauli spin matrices

$$\underline{\sigma}_1 = \begin{pmatrix} 0 & 1 \\ 1 & 0 \end{pmatrix}, \underline{\sigma}_2 = \begin{pmatrix} 0 & -i \\ i & 0 \end{pmatrix}, \underline{\sigma}_3 = \begin{pmatrix} 1 & 0 \\ 0 & -1 \end{pmatrix} \tag{9.12}$$

---

[7]An equivalent algebra with a different complex structure is formed by the quaternions (bivectors) taken over the complex field. The general complex quaternion is a complex linear combination of the basis elements $\{1, \mathbf{e}_1\mathbf{e}_2, \mathbf{e}_2\mathbf{e}_3, \mathbf{e}_3\mathbf{e}_1\}$ in $\mathcal{V}_0 \oplus \mathcal{V}_2$.

as long as the Kronecker delta δ_{jk} is replaced by $\delta_{jk}\mathbf{1}$ where $\mathbf{1}$ is the unit matrix. The matrix algebra of products and linear combinations of the spin matrices obeys the same product rule as the Clifford algebra. We say that the Pauli matrices provide a *matrix represention* of the algebra. Indeed, they give a lowest-dimensional faithful matrix representation of the algebra and give the algebra its name: the Pauli algebra. The full representation is realized simply by replacing the unit scalar by the unit matrix

$$\mathbf{1} \equiv \underline{\sigma}_0 = \begin{pmatrix} 1 & 0 \\ 0 & 1 \end{pmatrix} \tag{9.13}$$

and every unit vector $\mathbf{e}_k$ by the corresponding matrix $\underline{\sigma}_k$.

Exercise 9.2. Using the above matrix representation,

(a) prove that the rule (9.11) is obeyed;

(b) show that an arbitrary Pauli cliffor $p = p_0 + \mathbf{p}$ has the representation

$$\underline{p} = \begin{pmatrix} p^0 + p^3 & p_- \\ p_+ & p^0 - p^3 \end{pmatrix} \tag{9.14}$$

where $p_\pm = p^1 \pm ip^2$ and $p_0 = p^0$ is a scalar.

(c) Given the matrix representation $\underline{p} = p^\mu \underline{\sigma}_\mu$, show that the scalar and vector components of p are given by

$$p^\mu = \frac{1}{2} tr \left(\underline{p}\underline{\sigma}_\mu \right) \tag{9.15}$$

where tr is the trace[8] of the matrix.

(d) Noting that $p_+ p_- = (p^1)^2 + (p^2)^2$, demonstrate that the determinant of p is

$$\det \underline{p} = p_0^2 - \mathbf{p}^2. \tag{9.16}$$

Of course, many other representations are possible. The algebra is more general than any one of them, and by avoiding any specific

---

[8]The trace of a square matrix is the sum of the elements on its principal diagonal.

matrix representation, we can express the elements of the algebra in a component-free form, independent of any coordinate axes. Thus, our use of the term "Pauli algebra" refers to the algebraic structure, the geometric algebra of 3-dimensional Euclidean space, and not to any particular matrix representation.

9.3 Geometrical Significance

Any cliffor in the Pauli algebra can be written as the linear combination

$$p = p^\mu e_\mu + q^\nu i e_\nu = (p^\mu + i q^\mu) e_\mu, \qquad (9.17)$$

where the components p^μ, q^ν are real scalars, and for simplicity we have put $e_k = \mathbf{e}_k$, $k = 1, 2, 3$, and $e_0 = 1$. The association of the imaginary i with the handedness of the unit volume element $\mathbf{e}_1 \mathbf{e}_2 \mathbf{e}_3$ of 3-dimensional space lends geometrical meaning to imaginary numbers and vectors. We assume that $\mathbf{e}_1, \mathbf{e}_2$, and $\mathbf{e}_3$ form a right-handed basis, that is, turning the slot of a right-handed screw from $\mathbf{e}_1$ to $\mathbf{e}_2$ causes the screw to advance along the $\mathbf{e}_3$ direction. The imaginary i then represents a right handedness which relates a turning motion in a plane to linear motion along the normal to it. Like the right hand, the turning motion in the plane (indicated by the curling fingers of the right hand) and the linear motion along its normal (indicated by the direction of the thumb) can have any orientation in space, but the two motions are fixed relative to one another. Similarly, $-i$ signifies a left handedness.

The physical significance of (9.17) is clearer if we consider p to be the sum of a real scalar $p_0 = p^0$, a real vector $\mathbf{p} = p^k \mathbf{e}_k$, a bivector or *pseudovector* $i\mathbf{q} = iq^j \mathbf{e}_j$ (represented by an imaginary vector), and a trivector or *pseudoscalar* $iq_0 = iq^0$ (represented by an imaginary scalar). These four parts are distinguished by their behavior under rotation and spatial inversion. Rotation affects both real vectors and pseudovectors (imaginary vectors) in the same way and leaves real scalars and pseudoscalars (imaginary scalars) invariant. Spatial inversion, on the other hand, replaces the unit vectors $\mathbf{e}_j$ by $-\mathbf{e}_j$ and thus changes $i = \mathbf{e}_1 \mathbf{e}_2 \mathbf{e}_3$ to $-i$; it consequently changes right-handed coordinate systems into left-handed ones and reverses the sign of real vectors and imaginary scalars but not of real scalars and imaginary vectors.

Real scalars include common physical quantities like charge, mass, length, density, time, and energy, which are unchanged by either rotations or inversions.[9] Relative positions, momenta, velocities, forces, acceleration, and electric fields are examples of real vectors. An imaginary vector in $\mathcal{P}$ results if the handedness of i is divided by a vector which represents the direction of linear motion. It thus represents a turning motion in a plane, such as a plane of rotation. Cross products of vectors, such as angular momenta and magnetic fields (when expressed in terms of current sources), are the duals of pseudovectors: they point in a direction normal to the plane of rotation represented by the pseudovector. The dual vector to a plane of rotation thus lies along the axis of rotation, and we call it an *axial* vector.[10] Pseudoscalars arise as scalar products of pseudovectors with vectors. Although they should be imaginary scalars, it is usually the real dual quantity which is used in traditional treatments. For example, the dot product of the angular momentum of the spinning motion of a particle with the direction of its linear momentum gives the *helicity* of the particle, which is the dual of a pseudoscalar.

---

[9]However, note that length, density, time, and energy are not invariant under boosts and are therefore not *Lorentz* scalars (see Section 9.6).

[10]The terms *axial vector* and *pseudovector* are often treated as synonymous, but it is quite useful to be able to distinguish a bivector (an imaginary vector, the pseudovector) from its dual (a real vector, the axial vector). The distinction reveals a subtlety about their behavior under spatial inversion. If $\mathbf{a}$ and $\mathbf{b}$ are regular polar vectors such as position and velocity vectors, the bivector representing the plane containing them is the imaginary vector $(\mathbf{ab} - \mathbf{ba})/2 = i(\mathbf{a} \times \mathbf{b})$, which is also known as the exterior or Grassmann product $\mathbf{a} \wedge \mathbf{b}$. The axial vector is the real dual vector normal to the plane, namely $\mathbf{a} \times \mathbf{b} = a^j b^k \epsilon_{jkl} \mathbf{e}_l$. If the vectors are inverted $\mathbf{a} \rightarrow -\mathbf{a}$, $\mathbf{b} \rightarrow -\mathbf{b}$, by changing the signs of their components while leaving the basis vectors fixed, both the pseudovector and the axial vector are invariant. On the other hand, if the components are held fixed and the basis vectors $\mathbf{e}_k$ inverted, then the pseudovector is invariant, but the axial vector changes sign. This behavior of axial vectors is consistent with their definition as vector-like elements which, in contrast to polar vectors, change sign under a transformation from a right-handed coordinate system to a left-handed one [see, for example, R. Baierlein, *Newtonian Dynamics* (McGraw-Hill, New York 1983), p. 316]. Such a transformation can be realized by the simultaneous change in the signs of components (of polar vectors) and the directions of basis vectors.

9.4   Examples

Consider some simple examples.

9.4.1   Dot and cross products

Let $\mathbf{a}, \mathbf{b}$ be any two vectors. From the general multiplication rule (9.11) of the algebra, it follows that

$$
\begin{aligned}
\mathbf{ab} &= \mathbf{a} \cdot \mathbf{b} + i\mathbf{a} \times \mathbf{b} && (9.18) \\
&= |\mathbf{a}|\,|\mathbf{b}| \left(\cos\theta + i\hat{\boldsymbol{\theta}} \sin\theta \right) \\
&= |\mathbf{a}|\,|\mathbf{b}| \exp\left(i\theta \right)
\end{aligned}
$$

where θ is the angle from $\mathbf{a}$ to $\mathbf{b}$ and the direction $\hat{\boldsymbol{\theta}}$ is parallel to $\mathbf{a} \times \mathbf{b}$. The real scalar part of the product is the dot product, which is symmetric in the two factors $\mathbf{a}$ and $\mathbf{b}$:

$$
\mathbf{a} \cdot \mathbf{b} = \left(\mathbf{ab} + \mathbf{ba} \right)/2. \qquad (9.19)
$$

The imaginary vector part (the bivector or pseudovector part) gives the orientation of the plane formed by $\mathbf{a}$ and $\mathbf{b}$, and its magnitude is equal to the area of the parallelogram bordered by $\mathbf{a}$ and $\mathbf{b}$ (see Fig. 9.1). It is the antisymmetric part of the product $\mathbf{ab}$:

$$
i\mathbf{a} \times \mathbf{b} = \left(\mathbf{ab} - \mathbf{ba} \right)/2. \qquad (9.20)
$$

The vector dual to the plane is the cross product. From (9.18), parallel vectors commute whereas perpendicular ones anticommute:

$$
\begin{aligned}
\mathbf{a} \| \mathbf{b} &\iff \mathbf{ab} - \mathbf{ba} = 0 \\
\mathbf{a} \perp \mathbf{b} &\iff \mathbf{ab} + \mathbf{ba} = 0
\end{aligned} \qquad (9.21)
$$

Exercise 9.3. Evaluate and interpret the product of a unit vector along the x-axis with another unit vector in the xy-plane at an angle ϕ to the x-axis. Show that the product is unchanged if both unit vectors are rotated by an angle θ about $\mathbf{e}_3$.
Answer: The product is

$$
\mathbf{e}_1 \left(\mathbf{e}_1 \cos\phi + \mathbf{e}_2 \sin\phi \right) = \cos\phi + i\mathbf{e}_3 \sin\phi = \exp\left(i\mathbf{e}_3\phi \right),
$$

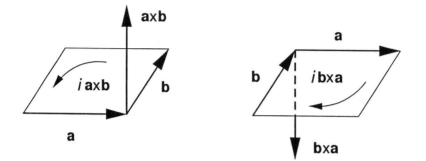

Figure 9.1. The bivectors $i\mathbf{a} \times \mathbf{b}$ and $i\mathbf{b} \times \mathbf{a} = -i\mathbf{a} \times \mathbf{b}$ represent plane areas of opposite orientation whose edges are parallel to the vectors $\mathbf{a}$ and $\mathbf{b}$. In three dimensions, they are also pseudovectors, and their duals are the axial vectors $\mathbf{a} \times \mathbf{b}$ and $\mathbf{b} \times \mathbf{a}$, respectively, normal to the plane. The orientation of the plane is related to the sense of circulation of $\mathbf{a}$ and $\mathbf{b}$ around the periphery.

which means that the projection of one unit vector on the other is $\cos\phi$ and that the unit vectors are the sides of a parallelogram of area $\sin\phi$ in the $i\mathbf{e}_3 = \mathbf{e}_1\mathbf{e}_2$ plane. Compare the product

$$(\mathbf{e}_1 \cos\theta + \mathbf{e}_2 \sin\theta)\,(\mathbf{e}_1 \cos(\theta + \phi) + \mathbf{e}_2 \sin(\theta + \phi))$$

and note that, consistent with relation (9.18), it is independent of θ.

Exercise 9.4. Evaluate the "triple cross product" in terms of dot products.
Solution: Use the relation $\mathbf{b} \times \mathbf{c} = (\mathbf{bc} - \mathbf{cb})/(2i)$ to expand the cross products, then add and subtract terms to

complete the dot-product terms, and finally collect like terms:

$$
\begin{aligned}
\mathbf{a} \times (\mathbf{b} \times \mathbf{c}) &= \tfrac{1}{4}(\mathbf{acb} + \mathbf{bca} - \mathbf{abc} - \mathbf{cba}) \\
&\quad + \tfrac{1}{4}(\mathbf{cab} + \mathbf{bac} - \mathbf{bac} - \mathbf{cab}) \\
&= [(\mathbf{a} \cdot \mathbf{c})\,\mathbf{b} + \mathbf{b}\,(\mathbf{a} \cdot \mathbf{c}) - (\mathbf{a} \cdot \mathbf{b})\,\mathbf{c} - \mathbf{c}\,(\mathbf{a} \cdot \mathbf{b})]\,/2 \\
&= (\mathbf{a} \cdot \mathbf{c})\,\mathbf{b} - (\mathbf{a} \cdot \mathbf{b})\,\mathbf{c}.
\end{aligned}
\tag{9.22}
$$

The parentheses around the dot products can be omitted if we agree that where dot products and algebraic products appear in the same level of parenthesis, the dot products are computed before the algebraic products. (See also the footnote at the end of Section 9.4.)

Exercise 9.5. Let $\mathbf{v}$ be an arbitrary real vector in three-dimensional space. Expand $\mathbf{v}$ in a Cartesian basis $\{\mathbf{e}_1, \mathbf{e}_2, \mathbf{e}_3\}$ and calculate $(i\mathbf{e}_1)\,\mathbf{v}\,(i\mathbf{e}_1)$ as a linear combination of the orthonormal basis vectors. Thereby show that it represents the vector $\mathbf{v}$ after reflection in the $i\mathbf{e}_1$ plane.

Exercise 9.6. Let $\mathbf{e}$ be a real unit vector. Expand the product $\mathbf{ve}$ in accordance with the first line of equation (9.18) to prove that any vector $\mathbf{v}$ can be written

$$
\mathbf{v} = \mathbf{vee} = (\mathbf{v} \cdot \mathbf{e})\,\mathbf{e} + \mathbf{e} \times (\mathbf{v} \times \mathbf{e}) \ . \tag{9.23}
$$

Interpret the result to show that any vector can be expressed as the sum of vectors parallel and perpendicular to an arbitrary direction $\mathbf{e}$.

9.4.2 Algebraic functions of vectors

The exponential expression in (9.18) is meaningful because in the vector algebra we can easily calculate any power of a vector, and thus any analytic function f of it:

$$
\mathbf{a}^n = \begin{cases} |\mathbf{a}|^n & n \text{ even} \\ |\mathbf{a}|^n\,\hat{\mathbf{a}} & n \text{ odd} \end{cases} \tag{9.24}
$$

$$
f(\mathbf{a}) = f_+(|\mathbf{a}|) + \hat{\mathbf{a}} f_-(|\mathbf{a}|) \tag{9.25}
$$

where $f_\pm(x) = \frac{1}{2}[f(x) \pm f(-x)]$ is the even or odd part of $f(x)$. Such a function is a cliffor of the algebra with, in general, both scalar and vector parts, both of which may be complex. The scalar part is the even part of the function, and the vector part, which is parallel to the vector argument, is the odd part.[11] Two important examples of analytic functions of vectors are the exponential functions

$$\exp(\mathbf{a}) = \cosh|\mathbf{a}| + \mathbf{a}\frac{\sinh|\mathbf{a}|}{|\mathbf{a}|} \tag{9.26}$$

$$\exp(i\mathbf{b}) = \cos|\mathbf{b}| + i\mathbf{b}\frac{\sin|\mathbf{b}|}{|\mathbf{b}|}. \tag{9.27}$$

Functions of vectors are often useful even when the function cannot be expanded in a series of positive powers. An example is the inverse $\mathbf{a}^{-1}$ of a vector. Since $\mathbf{a}^2$ is a scalar, the vector

$$\mathbf{a}^{-1} = \frac{\mathbf{a}}{\mathbf{a}^2} \tag{9.28}$$

is evidently the inverse of $\mathbf{a}$. As long as $\mathbf{a}^2 \neq 0$, relation (9.24) can be extended to negative integer powers.[12]

Exercise 9.7. Rewrite $(1 + \mathbf{a})^3$ as the sum of a scalar and a vector.
Answer: $(1 + 3\mathbf{a}^2) + \mathbf{a}(3 + \mathbf{a}^2)$.

Exercise 9.8. Use the power-series expansion of the exponential to prove that

$$\exp(a_0)\exp(\mathbf{a}) = \exp(a_0 + \mathbf{a}).$$

Solution: Because a_0 and $\mathbf{a}$ commute, the proof is as in equation (3.43), namely

$$\exp(a_0)\exp(\mathbf{a}) = \sum_{j=0}^{\infty}\sum_{k=0}^{\infty}\frac{a_0^j\,\mathbf{a}^k}{j!\,k!} = \sum_{n=0}^{\infty}\sum_{j=0}^{\infty}\binom{n}{j}\frac{a_0^j\mathbf{a}^{n-j}}{n!}$$

---

[11]Many scalar functions of vectors are not the scalar part of any algebraic function. An important example of such a scalar function is the linear function $\mathbf{r}\cdot\mathbf{a} = r^j a^j$, which defines an inner product of the argument $\mathbf{r}$ with a fixed vector $\mathbf{a}$.

[12]Functions of general Pauli elements can also be found. See Problems 2 and 3 at the end of this chapter.

$$= \sum_{n=0}^{\infty} \frac{(a_0 + \mathbf{a})^n}{n!} = \exp(a_0 + \mathbf{a}). \quad (9.29)$$

Exercise 9.9. Let $\mathbf{a}, \mathbf{b}$ be two real vectors. Prove that the equality $\exp(\mathbf{a} + \mathbf{b}) = \exp(\mathbf{a}) \exp(\mathbf{b})$ holds iff $\mathbf{a}$ and $\mathbf{b}$ are aligned parallel or antiparallel.

Solution: If $\mathbf{ab} = \mathbf{ba}$, then the power-series expansion of the exponential gives $\exp(\mathbf{a} + \mathbf{b}) = \exp(\mathbf{a}) \exp(\mathbf{b})$ as in the previous example. On the other hand, if $\exp(\mathbf{a} + \mathbf{b}) = \exp(\mathbf{a}) \exp(\mathbf{b})$, then an expansion of the right-hand side contains an imaginary vector part

$$i\mathbf{a} \times \mathbf{b} \frac{\sinh|\mathbf{a}| \sinh|\mathbf{b}|}{|\mathbf{a}| \, |\mathbf{b}|} \quad (9.30)$$

whereas the left-hand side is a power series in a real vector and is therefore real. Consequently, $\mathbf{a} \times \mathbf{b} = 0$, and hence $\mathbf{a}$ and $\mathbf{b}$ must be aligned.

9.4.3 Rotations in a plane

The physical significance of $\hat{\mathbf{a}}\hat{\mathbf{b}} = \exp(i\boldsymbol{\theta})$ becomes clear if we multiply it from the left by any vector $\mathbf{r}$ lying in the $i\boldsymbol{\theta}$ plane:

$$\begin{aligned} \mathbf{r} \exp(i\boldsymbol{\theta}) &= \exp(-i\boldsymbol{\theta})\,\mathbf{r} = \left(\cos\theta - i\hat{\boldsymbol{\theta}}\sin\theta\right)\mathbf{r} \\ &= \mathbf{r}\cos\theta + \hat{\boldsymbol{\theta}} \times \mathbf{r}\sin\theta. \end{aligned} \quad (9.31)$$

The result describes the vector $\mathbf{r}$ after a rotation: one can regard $\hat{\mathbf{a}}\hat{\mathbf{b}} = \exp(i\boldsymbol{\theta})$ as the operator which rotates any vector in the plane $i\boldsymbol{\theta}$ by the angle θ which separates $\hat{\mathbf{a}}$ and $\hat{\mathbf{b}}$ in the plane. As a special case, it rotates $\hat{\mathbf{a}}$ into $\hat{\mathbf{b}}$:

$$\hat{\mathbf{a}} \exp(i\boldsymbol{\theta}) = \hat{\mathbf{a}}\hat{\mathbf{a}}\hat{\mathbf{b}} = \hat{\mathbf{b}}, \quad (9.32)$$

but more generally, it rotates any vector in the plane of $\hat{\mathbf{a}}$ and $\hat{\mathbf{b}}$ by θ.

The axial vector dual to the plane of rotation $i\boldsymbol{\theta}$ is $\hat{\boldsymbol{\theta}}$, the axis of rotation. Successive rotations about the same axis $\hat{\boldsymbol{\theta}}$ commute and combine by adding the angles: $\exp\left(i\theta_1\hat{\boldsymbol{\theta}}\right)\exp\left(i\theta_2\hat{\boldsymbol{\theta}}\right) = \exp\left[i\left(\theta_1 + \theta_2\right)\hat{\boldsymbol{\theta}}\right]$. The usual trigonometric identities for the sine and cosine of a sum of angles is obtained by expanding both sides of this result as in (9.31). This algebraic formalism for rotating vectors in a plane should look familiar, since it is closely related to the Wessel-Argand diagrams of complex scalars, which is frequently used to describe such rotations (see Chapter 8). The precise relationship between the two formalisms will be established in the next section, but first we want to show how the algebraic formalism, because it contains information about the rotation axis and the orientation of the rotation plane, can be extended to 3-dimensional rotations.

9.4.4   Rotations in three dimensions

If the vector $\mathbf{r}$ does not lie in the rotation plane $i\boldsymbol{\theta}$, then we rotate only that part $\mathbf{r}_\perp = \mathbf{r} - \mathbf{r}_\parallel$ in the plane, that is, perpendicular to $\boldsymbol{\theta}$:

$$\mathbf{r} \to \mathbf{r}_\parallel + \mathbf{r}_\perp \exp\left(i\boldsymbol{\theta}\right). \tag{9.33}$$

Since perpendicular vectors anticommute [see (9.21)], the general rotation (9.33) can be put in the convenient form

$$\mathbf{r} \to \exp\left(-i\boldsymbol{\theta}/2\right) \mathbf{r} \exp\left(i\boldsymbol{\theta}/2\right). \tag{9.34}$$

Any 3-dimensional rotation can be represented by the elements $\pm \exp\left(-i\boldsymbol{\theta}/2\right)$.[13] The product of any two rotations is another rotation, whose angle and rotation axis can be found directly by expanding the exponentials as in (9.31). The noncommutivity of rotations about different axes is seen to be a result of the noncommutivity of vectors in different directions.

>   **Exercise 9.10.** Let the vector $\mathbf{r}$ be obtained by rotating $r\mathbf{e}_3$ first by an angle θ about the $\mathbf{e}_2$-axis and then by ϕ about the $\mathbf{e}_3$-axis. Find the three Cartesian coordinates of

---

[13]These elements form the group SU(2), which is the group of 2×2 matrices whose elements $\mathcal{R}$ are (1) *special* (that is, *unimodular*): $R\bar{R} = \bar{R}R = 1$; and (2) *unitary*: $RR^\dagger = 1 = R^\dagger R$. The name is also used for any group isomorphic to this group of matrices. (The bar and dagger (†) of an element is defined in Section 9.5.)

r in terms of r, θ, ϕ.

Solution: Writing the 3-dimensional rotations as above,

$$
\begin{aligned}
\mathbf{r} &= e^{-i\mathbf{e}_3\phi/2} \left(e^{-i\mathbf{e}_2\theta/2} r \mathbf{e}_3 e^{i\mathbf{e}_2\theta/2} \right) e^{i\mathbf{e}_3\phi/2} \\
&= r e^{-i\mathbf{e}_3\phi/2} \left(e^{-i\mathbf{e}_2\theta} \mathbf{e}_3 \right) e^{i\mathbf{e}_3\phi/2} \\
&= r \left(\mathbf{e}_1 e^{i\mathbf{e}_2\phi} \sin\theta + \mathbf{e}_3 \cos\theta \right) \quad (9.35) \\
&= r \left[(\mathbf{e}_1 \cos\phi + \mathbf{e}_2 \sin\phi) \sin\theta + \mathbf{e}_3 \cos\theta \right].
\end{aligned}
$$

The Cartesian coordinates are therefore

$$
\begin{aligned}
x &= r \sin\theta \cos\phi \\
y &= r \sin\theta \sin\phi \quad (9.36) \\
z &= r \cos\theta.
\end{aligned}
$$

9.4.5 Rotating frames

The algebraic formalism for rotations is well suited to handling transformations to and from rotating systems. Let **r** be any vector in a frame rotating at constant angular velocity $\boldsymbol{\omega}$. The corresponding vector **r**′ in the laboratory frame is

$$
\mathbf{r}' = \exp\left(-i\boldsymbol{\omega} t/2\right) \mathbf{r} \exp\left(i\boldsymbol{\omega} t/2\right). \quad (9.37)
$$

To study motion, we need its time derivative. Therefore note

$$
\frac{d}{dt} \exp\left(-i\boldsymbol{\omega} t/2\right) = \frac{-i\boldsymbol{\omega}}{2} \exp\left(-i\boldsymbol{\omega} t/2\right) \quad (9.38)
$$

as may be verified by expanding the exponential into even and odd parts: $\exp\left(-i\boldsymbol{\omega} t/2\right) = \cos\left(\omega t/2\right) - i\left(\hat{\boldsymbol{\omega}}/2\right) \sin\left(\omega t/2\right)$ and differentiating. Thus

$$
\begin{aligned}
\dot{\mathbf{r}}' &= \exp\left(-i\boldsymbol{\omega} t/2\right) \left(-\tfrac{i}{2}\left[\boldsymbol{\omega}\mathbf{r} - \mathbf{r}\boldsymbol{\omega}\right] + \dot{\mathbf{r}}\right) \exp\left(i\boldsymbol{\omega} t/2\right) \\
&= \exp\left(-i\boldsymbol{\omega} t/2\right) \left(\boldsymbol{\omega} \times \mathbf{r} + \dot{\mathbf{r}}\right) \exp\left(i\boldsymbol{\omega} t/2\right).
\end{aligned} \quad (9.39)
$$

Although this relation is traditionally written without the explicit rotation operators, their presence accurately describes the relation between the frames and is essential for unambiguous treatments. Higher-order derivatives are similarly calculated. The algebra of rotations can also give concrete descriptions of cycloid motion:

$$
\mathbf{r}(t) = \mathbf{V}(t) + \exp\left(-i\boldsymbol{\omega} t\right) \mathbf{r}_0, \quad (9.40)
$$

where the constant vectors $\mathbf{V}$ and $\mathbf{r}_0$ lie in the plane $i\boldsymbol{\omega}$. The term $\exp(-i\boldsymbol{\omega}t)\,\mathbf{r}_0$ describes uniform circular motion of radius $|\mathbf{r}_0|$ about the axis $\boldsymbol{\omega}$ whose position is $\mathbf{V}t$. Such motion, for example, solves the Lorentz-force equation for a charge in crossed electric and magnetic fields; $\mathbf{V}$ is then the drift velocity and $\boldsymbol{\omega}$ the angular cyclotron frequency.

Exercise 9.11. Show that the cycloid motion

$$\mathbf{r}(t) = \mathbf{r}(0) - \mathbf{r}_0 + \mathbf{V}t + \exp(-i\boldsymbol{\omega}t)\,\mathbf{r}_0 \qquad (9.41)$$

is the general solution to second-order differential equations of the form

$$\ddot{\mathbf{r}} = \mathbf{g} + \boldsymbol{\omega} \times \dot{\mathbf{r}} \qquad (9.42)$$

where $\mathbf{g}$ is a constant vector perpendicular to $\boldsymbol{\omega}$. Relate $\mathbf{g}, \mathbf{V}$, and $\boldsymbol{\omega}$.

Solution: Differentiation of $\mathbf{r}(t)$ gives

$$\dot{\mathbf{r}} = \mathbf{V} + \boldsymbol{\omega} \times \left(e^{-i\boldsymbol{\omega}t}\mathbf{r}_0\right)$$
$$\ddot{\mathbf{r}} = \boldsymbol{\omega} \times (\dot{\mathbf{r}} - \mathbf{V}) = \mathbf{g} + \boldsymbol{\omega} \times \dot{\mathbf{r}} \qquad (9.43)$$

where $\mathbf{g} = \mathbf{V} \times \boldsymbol{\omega}$. The solution can accommodate any initial position $\mathbf{r}(0)$ and velocity $\dot{\mathbf{r}}(0) = \mathbf{V} + \boldsymbol{\omega} \times \mathbf{r}_0$ and is therefore a general solution of the second-order equation of motion.

9.4.6   Reflection in planes

A vector $\mathbf{r}$ is reflected in a plane $i\hat{\mathbf{a}}$ by the simple transformation

$$\mathbf{r} \rightarrow (i\hat{\mathbf{a}})\,\mathbf{r}\,(i\hat{\mathbf{a}}) = -\hat{\mathbf{a}}\mathbf{r}\hat{\mathbf{a}} = \mathbf{r} - 2\mathbf{r} \cdot \hat{\mathbf{a}}\hat{\mathbf{a}}\,, \qquad (9.44)$$

which changes the sign of the component parallel to the unit normal $\hat{\mathbf{a}}$. The last equality follows directly from (9.21) when $\mathbf{r}$ is split into parts parallel and perpendicular to $\hat{\mathbf{a}}$. The result is the same for reflection in the plane $-i\hat{\mathbf{a}}$. In other words, the reflection is independent of the sense of rotation in the plane. Because $i\hat{\mathbf{a}} = \exp(i\hat{\mathbf{a}}\pi/2)$, the reflection (9.44) is seen to be equivalent to a 180-degree rotation in the plane $i\hat{\mathbf{a}}$ (see 9.34) together with an inversion $\mathbf{r} \rightarrow -\mathbf{r}$.

Two successive reflections in planes $i\hat{\mathbf{a}}$ and $i\hat{\mathbf{b}}$

$$\mathbf{r} \to i\hat{\mathbf{b}}\,(i\hat{\mathbf{a}})\,\mathbf{r}\,(i\hat{\mathbf{a}})\,i\hat{\mathbf{b}} = \left(\hat{\mathbf{b}}\hat{\mathbf{a}}\right)\mathbf{r}\left(\hat{\mathbf{a}}\hat{\mathbf{b}}\right)$$
$$= \exp\left(-i\theta\right)\mathbf{r}\exp\left(i\theta\right) \tag{9.45}$$

are equivalent to a rotation by 2θ, twice the angle between the two planes (see Sections 9.4.3 and 9.4.4), about an axis along the intersection of the planes. This effect is familiar from mirrors in the fitting area of clothing stores; images arising from double reflections show us, from various perspectives, as others see us.

The combination of three reflections in orthogonal planes is easily seen to be the same as spatial inversion. For example, successive reflections of $\mathbf{r}$ in the $i\mathbf{e}_1, i\mathbf{e}_2,$ and $i\mathbf{e}_3$ planes transforms $\mathbf{r}$ into

$$(i\mathbf{e}_3)\,(i\mathbf{e}_2)\,(i\mathbf{e}_1)\,\mathbf{r}\,(i\mathbf{e}_1)\,(i\mathbf{e}_2)\,(i\mathbf{e}_3) = -\mathbf{e}_3\mathbf{e}_2\mathbf{e}_1\mathbf{r}\mathbf{e}_1\mathbf{e}_2\mathbf{e}_3 = -\mathbf{r}, \tag{9.46}$$

since $\mathbf{e}_1\mathbf{e}_2\mathbf{e}_3 = i = -\mathbf{e}_3\mathbf{e}_2\mathbf{e}_1$. This is the principle of the corner-cube reflector, which by reflecting light in three orthogonal surfaces (three interior faces of a cube) always reflects a light beam back in the direction from which it came. (Molded plastic arrays of small corner cubes are used in common bicycle and motor-vehicle reflectors. Larger corner-cube arrays were placed on the surface of the moon to reflect laser light from the earth.)

> **Exercise 9.12.** Show that the effect of two consecutive reflections in the planes $i\hat{\mathbf{a}}$, $i\hat{\mathbf{b}}$ is invariant under any rotation of both planes about their line of intersection.
> **Solution:** It is sufficient to show that the product $\mathbf{ab}$ is invariant under a rotation in the plane $i\mathbf{a} \times \mathbf{b}$, the axial vector $\mathbf{a} \times \mathbf{b}$ of which is aligned with the intersection of $i\mathbf{a}$ and $i\mathbf{b}$.

9.4.7 Angular momentum

An important example illustrating the winding motion in planes is the orbital angular momentum $\mathbf{L}$, where $i\mathbf{L}$ is just the imaginary vector part of the product $\mathbf{rp}$:

$$i\mathbf{L} = (\mathbf{rp} - \mathbf{pr})/2 = i\mathbf{r} \times \mathbf{p}. \tag{9.47}$$

The real vector **L** is the axial vector normal to every vector in the orbital plane, and because of the geometrical significance of i, i**L** is the bivector representing the orbital plane itself. Its magnitude is the area of the parallelogram formed by **r** and **p**, and hence $2m$ times the rate at which the orbital area is swept out. Of course, it is just this relationship that connects the conservation of angular momentum to Kepler's second law.

The product $\mathbf{abc} = (\mathbf{ab})\,\mathbf{c} = \mathbf{a}\,(\mathbf{bc})$ of three vectors has both a real vector and a trivector part, and in three dimensions the trivector is an imaginary scalar whose magnitude is the volume of the parallelepiped with edges parallel to **a**, **b**, **c** (see Fig. 9.2).[14]

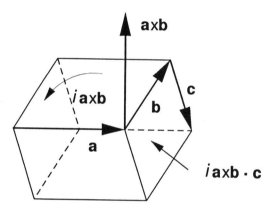

Figure 9.2. The oriented volume of the parallelepiped whose edges are parallel to vectors **a**, **b**, **c** is given by the trivector part of **abc**, and in three dimensions this trivector is the pseudoscalar $i\mathbf{a} \times \mathbf{b} \cdot \mathbf{c}$ whose magnitude gives the numerical size of the volume. In the figure, $\mathbf{a} \times \mathbf{b} \cdot \mathbf{c}$ is negative because $\mathbf{a} \times \mathbf{b}$ is roughly opposite to **c**.

Since the scalar part of a product of real or imaginary vectors is symmetric with respect to interchange of the two factors [see (9.19)],

---

[14]To keep the proliferation of parentheses in check, we impose the following priority rule: when dot, cross, and algebraic products appear within the same level of parenthesis, we perform the cross product first, then the dot product, and finally the algebraic product.

$(\mathbf{ab})\,\mathbf{c} = \mathbf{a}\,(\mathbf{bc})$, $\mathbf{c}\,(\mathbf{ab}) = (\mathbf{ca})\,\mathbf{b}$, and $\mathbf{b}\,(\mathbf{ca}) = (\mathbf{bc})\,\mathbf{a}$ all have the same scalar parts. The relations

$$\mathbf{a} \times \mathbf{b} \cdot \mathbf{c} = \mathbf{c} \times \mathbf{a} \cdot \mathbf{b} = \mathbf{b} \times \mathbf{c} \cdot \mathbf{a} = \mathbf{a} \cdot \mathbf{b} \times \mathbf{c} \qquad (9.48)$$

follow immediately. Note that in contrast to the field of real numbers, the algebra of vectors contains a noncommutative product (9.18) as well as some zero divisors (divisors or factors of zero): nonzero elements whose product vanishes, for example, $(1 + \mathbf{e}_3)\,(1 - \mathbf{e}_3) = 0$. Therefore the elements do not form a field: one must pay attention to the order of factors and be careful not to divide by a zero divisor. It is precisely these added complications which endow the Pauli algebra $\mathcal{P}$ with its rich mathematical structure capable of describing a wealth of physical phenomena.

9.5 The Structure of $\mathcal{P}$

So far, we have seen how complex scalars and vectors arise naturally in the vector algebra (also known as the Clifford algebra or geometric algebra) $\mathcal{P} = \mathcal{Cl}_3$ of real 3-dimensional space.[15] A general cliffor(9.17) of $\mathcal{P}$ may contain both scalars and vectors and may have both real and imaginary parts. The imaginary i carries important geometrical information in the algebra and is related to many occurrences of i in traditional formulations of physics. Some of the relations are most clearly seen if the structure of $\mathcal{P}$ is understood. Therefore, before we investigate further examples of $\mathcal{P}$ in relativity and electromagnetism, we pause to look at the shape, especially the subalgebras and "conjugations," of $\mathcal{P}$.

Since $\mathcal{P}$ includes the complex numbers, we expect to find a transformation in $\mathcal{P}$ corresponding to complex conjugation. In fact, $\mathcal{P}$ has a richer structure than the complex field and contains two principal "conjugations," which are more formally known as *involutions*. An involution I is a one-to-one mapping of $\mathcal{P}$ onto itself, $I : p \mapsto p'$; it is said to be *involutory* because when applied a second time, it returns

---

[15]To be more precise, the Pauli algebra $\mathcal{P}$ is $\mathcal{Cl}_3$ with the complexification which treats paravectors, and hence vectors of $\mathbb{R}^3$, as real. An alternative complexification would make all even elements of $\mathcal{Cl}_3$ real. Vectors of $\mathbb{R}^3$ would then be represented as i times a bivector and would thus be purely imaginary.

the elements to their original identities: $I : p' \mapsto p$. In other words, $I^2 = 1$. We call the involution corresponding to complex conjugation *Hermitean conjugation* and denote it by a dagger: $p \to p^\dagger$. The Hermitean conjugate of any element, written as a complex scalar plus a complex vector, is obtained by changing the sign of the imaginary i. Thus, for a general cliffor $p = (p^\mu + iq^\mu) e_\mu$ where p^μ and q^μ are real scalars, Hermitean conjugation is the transformation

$$p = (p^\mu + iq^\mu) e_\mu \to p^\dagger = (p^\mu + iq^\mu)^* e_\mu = (p^\mu - iq^\mu) e_\mu . \qquad (9.49)$$

The algebra $\mathcal{P}$ was generated from products of vectors in real three-dimensional space, and in terms of such products, Hermitean conjugation is known as *reversal*, since it is effected by reversing the order of multiplication of all real vectors. In particular, note the change of sign in the pseudoscalar element

$$i = e_1 e_2 e_3 \to (e_1 e_2 e_3)^\dagger = e_3 e_2 e_1 = -i . \qquad (9.50)$$

The other principal involution is *spatial reversal*[16]; it changes the sign of the complex vector part and is denoted by a bar: $p \to \bar{p}$. Thus, for an arbitrary cliffor p, written as the sum of a complex scalar $p_0 = p^0$ and a complex vector $\mathbf{p}$, spatial reversal is

$$p = p_0 + \mathbf{p} \to \bar{p} = p_0 - \mathbf{p} . \qquad (9.51)$$

Both Hermitean conjugation and spatial reversal reverse the order of multiplication in a product:

$$(ab)^\dagger = b^\dagger a^\dagger, \ \overline{(ab)} = \bar{b}\bar{a} , \qquad (9.52)$$

as may be shown by expanding the elements, multiplying them, and collecting terms into real and imaginary scalar and vector parts. When both spatial reversal and Hermitean conjugation are performed on an element, the result is *spatial inversion* $p \to \bar{p}^\dagger$, mentioned in the Section 9.3. Classical physics appears to be equally well-modeled by the Pauli algebra $\mathcal{P}$ and by its conjugates, $\bar{\mathcal{P}}, \mathcal{P}^\dagger$,and $\bar{\mathcal{P}}^\dagger$. Table 9.1 defines

---

[16]In th Clifford-algebra literature, this involution is sometimes called *Clifford conjugation*.

Operation	Action	Summary
Identity	$p \to p = p_0 + \mathbf{p} + i\mathbf{q} + iq_0$	$+ + ++$
Spatial reversal	$p \to \bar{p} = p_0 - \mathbf{p} - i\mathbf{q} + iq_0$	$+ - -+$
Hermitean conjugation	$p \to p^\dagger = p_0 + \mathbf{p} - i\mathbf{q} - iq_0$	$+ + --$
Spatial inversion	$p \to \bar{p}^\dagger = p_0 - \mathbf{p} + i\mathbf{q} - iq_0$	$+ - +-$

Table 9.1. The basic involutory transformations of $\mathcal{P}$

the action of these transformations on an arbitrary element and summarizes the results by specifying the relative signs of the real scalar, real vector, bivector (imaginary vector, pseudovector), and trivector (imaginary scalar, pseudoscalar) parts of the transformed elements.[17]

A cliffor p is said to be *even* iff $p = \bar{p}^\dagger$ and *odd* iff $p = -\bar{p}^\dagger$; *even* and *odd* are two *grades* of the algebra, and spatial inversion, by which even and odd elements are distinguished, is also known as the *grade automorphism*. An element is a (complex) *scalar*[18] iff $p = \bar{p}$ and a (complex) *vector* iff $p = -\bar{p}$; and it is *real* iff $p = p^\dagger$ and *imaginary* iff $p = -p^\dagger$. An arbitrary element p can always be split into real ($\Re$) and imaginary ($\Im$) parts, scalar (S) and vector (V) parts, and even ($+$) and odd ($-$) parts:

$$p = \frac{1}{2}\left(p + p^\dagger\right) + \frac{1}{2}\left(p - p^\dagger\right) = \langle p \rangle_\Re + \langle p \rangle_\Im \qquad (9.53)$$

$$p = \frac{1}{2}\left(p + \bar{p}\right) + \frac{1}{2}\left(p - \bar{p}\right) = \langle p \rangle_S + \langle p \rangle_V \qquad (9.54)$$

$$p = \frac{1}{2}\left(p + \bar{p}^\dagger\right) + \frac{1}{2}\left(p - \bar{p}^\dagger\right) = \langle p \rangle_+ + \langle p \rangle_- . \qquad (9.55)$$

The matrix representation of any real cliffor is Hermitean provided the basis vectors are given Hermitean representations. The matrix elements

---

[17]As long as $\mathcal{P}$ is thought of as a real algebra with eight basis forms, spatial inversion is an automorphism; it preserves the order of the products: $\left(\overline{ab}\right)^\dagger = \bar{a}^\dagger \bar{b}^\dagger$. Hermitean conjugation and spatial reversal, on the other hand, are involutory anti-automorphisms, or more simply, involutions, because they reverse the order of multiplication.

[18]While it is sometimes useful to distinguish between a (real) vector and a pseudovector and between a (real) scalar and a pseudoscalar, we find it more often convenient to use the terms "scalar" and "vector" in the complex sense: they can generally have both real and imaginary parts.

themselves are not necessarily real. Note that the product $pp^\dagger$ is always real, whereas the product $p\bar{p}$ is always a scalar. From the effects of the conjugations on the elements, it is seen that when conjugations are combined, the order of their application is immaterial:

$$(\bar{p})^\dagger = \overline{(p^\dagger)} = \bar{p}^\dagger . \tag{9.56}$$

There are several important subalgebras of the Pauli algebra $\mathcal{P}$. For example, a subalgebra is generated from products of any subset of the basis vectors $\mathbf{e}_1, \mathbf{e}_2, \mathbf{e}_3$. For example, the basis $\{\mathbf{e}_1, \mathbf{e}_2\}$ generates a Clifford algebra $\mathcal{C}\ell_2$ whose elements are linear combinations of $\{1, \mathbf{e}_1, \mathbf{e}_2, \mathbf{e}_1\mathbf{e}_2\}$. Another subalgebra is the *center* of $\mathcal{P}$, that is, the part of $\mathcal{P}$ which commutes with all elements. The center of $\mathcal{P}$ is just the scalar part $\mathcal{P}_S$ and comprises the complex field. The subalgebra $\mathcal{P}_S$ is a *field* because all its elements commute and there are no zero divisors. Still another subalgebra is the even Pauli algebra $\mathcal{P}_+$, which consists of all real linear combinations of real scalars and imaginary vectors of $\mathcal{P}$. Elements of $\mathcal{P}_+$ represent rotation-dilations, that is, rotations times a real scalar. Since the product of two imaginary vectors is a linear combination of a real scalar and an imaginary vector, $\mathcal{P}_+$ is closed under multiplication as well as under addition and thus does indeed form a subalgebra of $\mathcal{P}$. It is equivalent ("isomorphic") to Hamilton's quaternion algebra $\mathbb{H}$, which is a division algebra because it contains no zero divisors. However, it is not a field since its elements do not necessarily commute. The even subalgebra $\mathbb{H}_+$ of $\mathbb{H}$ is spanned by $\{1, \mathbf{e}_2\mathbf{e}_1\}$ and is isomorphic to the complex numbers $\mathcal{P}_S$ as well as to the even subalgebra of $\mathcal{C}\ell_2$. Any real vector in the $\mathbf{e}_1\mathbf{e}_2$ plane can be mapped onto the complex numbers by multiplying it by the vector by $\mathbf{e}_1$ or any other unit vector in $\mathbf{e}_1\mathbf{e}_2$. Thus

$$\mathbf{r} = x\mathbf{e}_1 + y\mathbf{e}_2 \mapsto \mathbf{e}_1\mathbf{r} = x + y\mathbf{e}_1\mathbf{e}_2 \mapsto z = x + iy \tag{9.57}$$

establishes the formal relation between real vectors in a plane and complex numbers. The square length of the vector $\mathbf{r}$ is $\mathbf{r}^2 = (\mathbf{r}\mathbf{e}_1)(\mathbf{e}_1\mathbf{r}) = (\mathbf{e}_1\mathbf{r})^\dagger(\mathbf{e}_1\mathbf{r}) \mapsto z^\dagger z$, which relates it to square complex modulus, and the dot product of two two-dimensional vectors is given by

$$2\mathbf{r} \cdot \mathbf{r}' = (\mathbf{e}_1\mathbf{r}')^\dagger(\mathbf{e}_1\mathbf{r}) + (\mathbf{e}_1\mathbf{r})^\dagger(\mathbf{e}_1\mathbf{r}') \to 2 < z^\dagger z' >_\Re . \tag{9.58}$$

Of course, $\mathcal{P}$ has the advantage of giving a coordinate-free represen-tation of positions and rotations in three dimensions. Because of the correspondence between positions in a plane and the field of complex numbers, uniform circular motion (see Section 9.4.5) is described in the complex field by the phase factor $\exp(-i\omega t)$ as in Section 8.8. Since the simple harmonic motion of oscillators is the projection of uniform circular motion, it is only a small step to the use of such factors in oscillating circuits. Including an exponential decay $\exp(-\gamma t)$ of the amplitude leads one to speak of "complex (angular) frequencies" $\omega - i\gamma$.

Knowing how to form inverses is important in any algebra. The inverse c^{-1} of any nonzero complex element c of $\mathcal{P}_S$ is identified by noting that the product $cc^{\dagger}$ is a (real) scalar and that $cc^{\dagger} = c^{\dagger}c$. It follows that $c^{-1} = c^{\dagger}/(cc^{\dagger})$. The same relations hold for quaternions in $\mathcal{P}_{+}$. Inverses of cliffors in $\mathcal{P}$ are discussed in the next section.

9.6 Special Relativity

In special relativity, the spacetime 4-vector[19] or paravector is a basic ge-ometric (covariant) entity, and products of such paravectors give other covariant quantities. Many 3-dimensional vectors, such as position and momentum, arise as the spatial parts of paravectors in spacetime. Oth-ers, like angular momentum and electric and magnetic fields, result from products of 4-vectors. We seek a formulation of relativity which treats spacetime vectors and their products as single entities and which can avoid, as the Pauli algebra does for vectors in $\mathbb{R}^3$, cumbersome component notation. The notation should usually be "covariant," that is, most relations should be true in any inertial frame, and the math-ematical structure should emphasize the geometric properties of the quantities. However, the notation should also be closely tied to physi-cal measurements. Although space and time components are mixed by Lorentz transformations, the time dimension is qualitatively distinct from spatial dimensions in any given frame: no observer would confuse

---

[19]The expression "4-vector" is used here in its common meaning as a 4-dimensional vector and *not* in the Clifford-algebra sense of an exterior product of four vectors. Because of the potential for confusion, we prefer the terms "par-avector" and "spacetime vector" over the older "4-vector."

the two in measurements performed in the laboratory. Ideally our notation, while covariant, should allow a qualitative distinction between time and space in any given frame. We shall see below how these apparently contradictory goals of covariance, on the one hand, and of the distinction of space and time in any given frame, on the other, can be reconciled in the Pauli algebra.[20]

An obvious way to extend the advantages of a multivector algebra to special relativity is to use the geometric algebra based on Minkowski spacetime instead of on $\mathbb{R}^3$. Considerable work along these lines has been published by Hestenes, Salingaros, and others, and the geometric algebra of spacetime has a well-known representation based on 4×4 matrices.[21] This multivector algebra, sometimes known also as the real Dirac algebra, has the somewhat awkward feature that the volume element of the algebra anticommutes with vectors and pseudovectors and so is obviously distinct from the usual i.

However, as surprising as it may seem, there is no need to leave the relatively simple Pauli algebra in order to handle problems covariantly in relativity. We saw in Section 9.2 that the algebra of real 3-dimensional vectors generates a complex 4-dimensional space. Now we shall show that the Minkowski spacetime metric arises naturally in this space when one finds the multiplicative inverse of a general Pauli element.

An cliffor p has an inverse if there exists another element, say p', whose product with p is a nonvanishing complex scalar: $pp' = \overline{pp'} = \overline{p'}\bar{p}$. Indeed, the inverse is then simply $p^{-1} = p'/(pp')$. The square of any real vector $\mathbf{p}$ is a scalar, so that, as seen in Section 9.4.2, the inverse of $\mathbf{p}$ is $\mathbf{p}^{-1} = \mathbf{p}/(\mathbf{p}^2)$. However, the square of a general element p does not necessarily belong to $\mathcal{P}_S$. The product $p\bar{p}$ is always its own spatial

---

[20]W. E. Baylis, "Special relativity with 2×2 matrices," *Am. J. Phys.* **48**, 918–925 (1980); W. E. Baylis and G. Jones, "Special relativity with Clifford algebras and 2×2 matrices, and the exact product of two boosts," *J. Math. Phys.* **29**, 57–62 (1988); W. E. Baylis and G. Jones, "The Pauli algebra approach to special relativity," *J. Phys. A: Math. Gen.* **22**, 1–16 (1989).

[21]These matrices, which represent the basis vectors of spacetime, are the Dirac γ matrices. See, for example, N. Salingaros and M. Dresden, "Physical algebras in four dimensions. I. The Clifford algebra in Minkowski spacetime," *Adv. in Appl. Math.* **4**, 1–30 (1983).

reverse and is therefore a complex scalar. As long as $p\bar{p}$ doesn't vanish, p has the inverse

$$p^{-1} = \bar{p}/\left(p\bar{p}\right). \tag{9.59}$$

Evidently the (complex) scalar $p\bar{p}$ plays the same role for an element p of the complex four-dimensional paravector space of $\mathcal{P}$ that the scalar $\mathbf{r}^2$ plays for a vector $\mathbf{r}$ in real three-dimensional space: we will refer to $p\bar{p}$ as the square norm or square modulus of p. If the paravector $p = p_0 + \mathbf{p}$ is identified with the real spacetime vector $(p_0, \mathbf{p})$, its square norm

$$p\bar{p} = (p_0 + \mathbf{p})(p_0 - \mathbf{p}) = p_0^2 - \mathbf{p}^2 \tag{9.60}$$

is also real, and its form gives the Minkowski metric of spacetime. The scalar part of p is the zero or "time" component of the 4-vector (here we use units with the speed of light $c = 1$). Those paravectors with a vanishing norm are said to be light-like: $p_0^2 - \mathbf{p}^2 = 0$. They are seen to be zero divisors and hence noninvertible elements of the algebra.

Linear transformations which leave $p\bar{p}$ invariant for any spacetime paravector p are called Lorentz transformations. A Lorentz transformation L is said to be *proper* if its square modulus is 1. Rotations and boosts (*i.e.*, velocity transformations) are examples of proper Lorentz transformations; rotations and boosts (velocity changes) are examples. Spatial inversion is an example of an improper Lorentz transformation. Any scalar that is invariant under Lorentz transformations is said to be a *Lorentz scalar*.

Lorentz transformations which can be realized physically by a sequence of infinitesimal transformations are called *restricted Lorentz transformations* (proper and orthochronous); they are represented by a simple generalization of the rotations (9.34). Using

$$L = \exp\left(\mathbf{w}/2\right)\exp\left(-i\boldsymbol{\theta}/2\right), \tag{9.61}$$

the restricted Lorentz transformation of any paravector p is

$$p \rightarrow LpL^{\dagger} \tag{9.62}$$

where L is any unimodular element: $L\bar{L} = 1$, which can always be written as the product (9.61).[22] When $\boldsymbol{\theta} = 0$, then $L = \exp{(\mathbf{w}/2)}$ and is real; its application (9.62) gives a boost with the *boost parameter* $\mathbf{w}$. When $\mathbf{w} = 0$, L rotates the paravector by the angle θ about the axis $\hat{\boldsymbol{\theta}}$. As before (see 9.34), the rotation plane is $i\hat{\boldsymbol{\theta}}$. We thus see the familiar relation that a boost takes the form of an imaginary rotation, but now there is added physical meaning to the fact that the rotation parameter $i\hat{\boldsymbol{\theta}}$ is a bivector whereas the boost parameter $\mathbf{w}$ is a vector.

Lorentz transformations from the rest frame of an object to the laboratory frame (in which the object is moving) allow any physical quantity known in the rest frame to be determined as observed in the laboratory. As a simple example, consider the transformation of the paravector velocity (4-velocity) u from its rest-frame value $u_{rest} = 1$. The paravector velocity of an object is a time derivative of its spacetime position, which is the collection of its coordinates in the form $r = t + \mathbf{r}$. In the rest frame, the spatial part $\mathbf{r}$ of the spacetime position r is constant. The time derivative of r is thus $u_{rest} = 1$. Using the product form (9.61) of L, we obtain the paravector velocity in the laboratory frame:

$$u = Lu_{rest}L^{\dagger} = e^{\mathbf{w}}. \qquad (9.63)$$

On the other hand, if we boost an infinitesimal displacement $d\tau$ of the object in the rest frame to the laboratory frame, we obtain the corresponding displacement dr in the laboratory:

$$dr = Ld\tau L^{\dagger} = ud\tau. \qquad (9.64)$$

The changing part of the spacetime position in the rest frame is τ, a scalar element of the algebra. It is the time coordinate in the rest frame and is known as the *proper time* of the object. The paravector velocity is seen to be the proper-time rate of change of the spacetime position

$$u = \gamma + \mathbf{u} = \frac{dr}{d\tau} = \frac{dt}{d\tau} + \frac{d\mathbf{r}}{d\tau}, \qquad (9.65)$$

---

[22]In fact, any element can be expressed as the product of a real element and a unitary element. The group of unimodular elements is SL(2,C), the double covering group for restricted Lorentz transformations. Its unitary subgroup SU(2) covers rotations. The group SL(2,C) is said to be a six-parameter group, since every Lorentz transformation is uniquely determined by the choice of six parameters: the three real components of $\mathbf{w}$ and the three imaginary components of the $i\boldsymbol{\theta}$.

and we have used the usual notation for what is known as the *Lorentz (dilation) factor*

$$\gamma = \frac{dt}{d\tau}.\qquad(9.66)$$

The vector part **u** of the 4-velocity is related to the coordinate velocity $\mathbf{v} = d\mathbf{r}/dt$ by

$$\mathbf{u} = \frac{d\mathbf{r}}{d\tau} = \frac{dt}{d\tau}\frac{d\mathbf{r}}{dt} = \gamma\mathbf{v}.\qquad(9.67)$$

The scalar and vector parts of relation (9.63) provide the usual relations between the boost parameter $\mathbf{w} = w\hat{\mathbf{u}}$ and the 4-velocity u achieved by the boost,

$$\gamma = \cosh w, \quad \mathbf{u} = \hat{\mathbf{u}}\sinh w,\qquad(9.68)$$

and ensure the unimodularity of u:

$$u\bar{u} = \gamma^2\left(1 - \mathbf{v}^2\right) = 1.\qquad(9.69)$$

The resultant 4-velocity is independent of the initial rotation angle and can be identified with the square of the boost element $\exp\left(\mathbf{w}/2\right)$.

The application of two successive boosts to $u_{rest} = 1$, or equivalently, the boost of a 4-velocity, gives the rule for velocity composition:

$$U = \exp\left(\mathbf{w}'/2\right)u\exp\left(\mathbf{w}'/2\right) = u'\left(\gamma + \mathbf{u}_\parallel\right) + \mathbf{u}_\perp\qquad(9.70)$$

where $\mathbf{u}_\parallel$ and $\mathbf{u}_\perp$ are the parts of **u** parallel and perpendicular to $\mathbf{w}'$ and hence to u', respectively. The result is particularly simple when the boost parameters **w** and $\mathbf{w}'$ are parallel:

$$U = uu', \quad \mathbf{u} \times \mathbf{u}' = 0.\qquad(9.71)$$

The composition of parallel 4-velocities is thus given not by their sum but by their product. We can expand the relation into scalar and vector parts

$$U_0 =: \Gamma = \gamma'\gamma\left(1 + \mathbf{v}' \cdot \mathbf{v}\right), \quad \mathbf{U} =: \Gamma\mathbf{V} = \gamma'\gamma\left(\mathbf{v}' + \mathbf{v}\right)\qquad(9.72)$$

whose ratio gives the traditional form of the relationship:

$$\mathbf{V} = \frac{\mathbf{v} + \mathbf{v}'}{1 + \mathbf{v} \cdot \mathbf{v}'}, \quad \mathbf{v} \times \mathbf{v}' = 0.\qquad(9.73)$$

It is not much more difficult to expand the non-collinear case (9.70). Since the product of real elements is real only if the elements commute, the composition of boosts is a pure boost only if the boost directions are collinear. Otherwise, the product is the combination of a boost and a rotation. The rotation part is responsible for Thomas precession.

The 4-vectors, like p or u, transform as covariant elements as in equation (9.62) but may, in any given frame, be expanded into scalar (time) and vector (spatial) parts. The distinction between scalars and vectors mirrors the qualitative difference between time and space components. In this way, the Pauli algebra is able to accommodate both covariance and the distinctiveness of space and time components in any given frame.

> **Exercise 9.13.** Galaxies A and B are receding from the earth in opposite directions at 0.9 the speed of light. How fast is one moving with respect to the other? How can it be less than 1 (the speed of light)? Solution: Let $\mathbf{v}$ $(-\mathbf{v})$ be the velocity of galaxy A (B) as seen from earth. The velocity of earth as viewed from galaxy B is then $\mathbf{v}' = \mathbf{v}$. The corresponding 4-velocities are $u = u' = \gamma(1 + \mathbf{v})$ where from the discussion between (9.62) and (9.63), $\gamma = (1 + \mathbf{u}^2)^{1/2} = (1 - \mathbf{v}^2)^{-1/2}$. The 4-velocity of galaxy A as viewed from B is therefore
>
> $$U = u'u = \gamma^2\left(1 + \mathbf{v}^2 + 2\mathbf{v}\right)$$
>
> so that the velocity is
>
> $$\mathbf{V} = \frac{U}{\Gamma} = \frac{2\mathbf{v}}{1 + \mathbf{v}^2},$$
>
> and its magnitude is indeed less than 1:
>
> $$V = \frac{1.8}{1.81} = 0.994.$$

9.7   Electromagnetic Theory

The transformation properties of *products* of paravectors are easily found from the paravector transformation (9.62) and the unimodularity of L. An important example is $\partial \bar{A}$ where $\partial = \partial/\partial t - \nabla$ is the paravector (spacetime) *gradient operator* and $A = \phi + \mathbf{A}$ is the paravector potential. Let's first define a generalization of the scalar (dot) product of two vectors. If $a = a^\mu \mathbf{e}_\mu = a_0 + \mathbf{a}$ and $b = b^\nu \mathbf{e}_\nu = b_0 + \mathbf{b}$ are any two cliffors where the components a^μ, b^ν may be complex, then we define the dot product to be simply the complex scalar part of the product ab :

$$a \cdot b = \langle ab \rangle_S = \frac{1}{2} \left(ab + \overline{ab} \right) = a_0 b_0 + \mathbf{a} \cdot \mathbf{b} = \langle ba \rangle_S . \qquad (9.74)$$

Products of paravectors transform simply if paravectors are alternated with spatially reversed paravectors. The product $\partial \bar{A}$ transforms under restricted Lorentz transformations as

$$\partial \bar{A} \rightarrow L \partial L^\dagger \bar{L}^\dagger \bar{A} \bar{L} = L \partial \bar{A} \bar{L} \qquad (9.75)$$

where $L \bar{L} = 1 = \bar{L} L = \left(\bar{L} L \right)^\dagger = L^\dagger \bar{L}^\dagger$ (see Sections 9.5 and 9.6). Its scalar part is a Lorentz scalar:

$$\left\langle \partial \bar{A} \right\rangle_S \rightarrow \left\langle L \partial \bar{A} \bar{L} \right\rangle_S = \left\langle \bar{L} L \partial \bar{A} \right\rangle_S = \left\langle \partial \bar{A} \right\rangle_S. \qquad (9.76)$$

The remaining vector plus bivector together may be called a *parabivector* or a *6-vector*,[23] and although it transforms like any vector under rotations, its characteristic boost transformation is seen to be distinct.

The scalar, vector, pseudovector (bivector), and pseudoscalar (trivector) forms of $\mathcal{P}$, which belong to the four subspaces $\mathcal{V}_0, \mathcal{V}_1, \mathcal{V}_2, \mathcal{V}_3$, respectively, can thus be combined to make various covariant quantities:

Lorentz scalar	$\in \mathcal{V}_0$
paravector (4-vector)	$\in \mathcal{V}_0 \oplus \mathcal{V}_1$
parabivector (6-vector)	$\in \mathcal{V}_1 \oplus \mathcal{V}_2$
pseudo-paravector	$\in \mathcal{V}_2 \oplus \mathcal{V}_3$
Lorentz pseudoscalar	$\in \mathcal{V}_3$.

The product of two spacetime vectors (paravectors) is generally a Lorentz scalar plus a spacetime bivector (a parabivector). Similarly,

---

[23]The term "6-vector" refers to six components, three real vector components plus three imaginary vector (bivector) ones. It can also be identified as a spacetime bivector.

the product of three spacetime vectors or of a spacetime vector and a spacetime bivector can be shown to be a spacetime vector plus a spacetime pseudo-vector, whereas the product of two spacetime bivectors is a complex Lorentz scalar plus another spacetime bivector.

By writing out the individual terms, one sees that the vector part of $\partial \bar{A}$ gives the electromagnetic field:

$$\left\langle \partial \bar{A} \right\rangle_V = -\left(\frac{\partial \mathbf{A}}{\partial t} + \nabla \phi \right) + i\nabla \times \mathbf{A} = \mathbf{E} + i\mathbf{B} \equiv \mathbf{F} . \tag{9.77}$$

The electric field $\mathbf{E}$ is a vector, but the magnetic field $i\mathbf{B}$ enters as a pseudovector, emphasizing the role of circulation in the plane $i\mathbf{B}$ perpendicular to $\mathbf{B}$. Together, they form the covariant parabivector $\mathbf{F} = \mathbf{E} + i\mathbf{B}$. The real and imaginary scalar and vector parts of the single field equation

$$\bar{\partial} \mathbf{F} = 4\pi K \bar{j} \tag{9.78}$$

are exactly Maxwell's usual four microscopic equations:

$$\bar{\partial} \mathbf{F} = \left(\frac{\partial}{\partial t} + \nabla \right) (\mathbf{E} + i\mathbf{B}) = \tag{9.79}$$

$$\nabla \cdot \mathbf{E} + \left(\frac{\partial \mathbf{E}}{\partial t} - \nabla \times \mathbf{B} \right) + i\left(\frac{\partial \mathbf{B}}{\partial t} + \nabla \times \mathbf{E} \right) + i\nabla \cdot \mathbf{B} = 4\pi K \left(\rho - \mathbf{j} \right) .$$

Here j is the current density and K is a constant depending on units ($= 1$ in Gaussian units, $= 1/4\pi\epsilon_0$ in SI units, and $= 1/4\pi$ in Heaviside-Lorentz units). Magnetic monopoles can be added by making j and A complex.[24]

The general directed plane-wave solution to the wave equation (9.78) in source-free space ($j = 0$) is[25]

$$\mathbf{F}(x) = k\hat{\boldsymbol{\xi}} f\left(\bar{k} \cdot x \right) \tag{9.80}$$

---

[24] Jiansu Wei and W. E. Baylis, "Monopoles without strings: a conflict between the one-photon condition and duality invariance," *Found. Phys. Lett.* **4**, 537–556 (1991).

[25] W. E. Baylis and G. Jones, "Relativistic dynamics of charges in external fields: the Pauli algebra approach," *J. Phys. A: Math. Gen.* **22**, 17–29 (1989).

where $x = t + \mathbf{x}$ is the spacetime position, $k = \omega + \mathbf{k}$ is a constant propagation paravector which satisfies $k\bar{k} = \omega^2 - \mathbf{k}^2 = 0$ and $\bar{k} \cdot x = \omega t - \mathbf{k} \cdot \mathbf{x}$, $\hat{\boldsymbol{\xi}}$ is any unit vector in the plane perpendicular to $\mathbf{k}$, and f is any scalar function. Rotations of $\mathbf{F}$ about $\mathbf{k}$ are equivalent to multiplications by a phase factor:

$$\mathbf{F} \rightarrow \exp\left(-i\phi\hat{\mathbf{k}}/2\right) \mathbf{F} \exp\left(i\phi\hat{\mathbf{k}}/2\right)$$

$$= \exp\left(-i\phi\hat{\mathbf{k}}\right) \mathbf{F} = \exp\left(-i\phi\right)\mathbf{F} \tag{9.81}$$

where we noted that $\bar{k}\mathbf{F} = \omega\left(1 - \hat{\mathbf{k}}\right)\mathbf{F} = 0$ and therefore $\hat{\mathbf{k}}\mathbf{F} = \mathbf{F}$. Thus if the function $f\left(\bar{k} \cdot x\right)$ varies as $\exp\left(\mp i\bar{k} \cdot x\right)$, then at fixed $\mathbf{x}$ the phase of $\mathbf{F}$ advances at the rate $\pm\omega$, which is equivalent to a rotation at the angular rate ω about the direction $\pm\hat{\mathbf{k}}$. The plane-wave solution then represents a circularly polarized wave. On the other hand, if the phase of $f\left(\bar{k} \cdot x\right)$ is constant, the solution is plane polarized.

A similar complex form is familiar to most physicists; one usually takes the real part to represent the electric field $\mathbf{E}$ of a circularly-polarized plane wave. The phase of $f\left(\bar{k} \cdot x\right)$ then gives the direction of $\mathbf{E}$ in the plane $i\mathbf{k}$. In $\mathcal{P}$, the polarization is not necessarily circular, and the real and imaginary parts of $\mathbf{F}$ have definite meaning: the real part is a vector and hence the electric field, and the imaginary part is a pseudovector identified with i times the magnetic field.

9.8   *Maple Calculations

A collection of procedures has been written for computations with Maple in the Pauli algebra and is available in the directory 'tmlib' on the disk supplied with the text. As suggested in Chapter 1, the entire 'tmlib' directory can be copied to the 'maplev*n*' directory on your hard disk. You need to open 'tmlib' and make it your current working directory in order to load the worksheets. To load the table of Pauli procedures into your Maple session, you can then type

> `read 'Pauli.tab';`

The Pauli-algebra package can then be read in:

> with(Palg);

You should see a list of available procedures. Remember that in Maple you must use forward slashes to separate directories and that the entire path + filename must be placed in back quotes. If the directories have been separated, or if you cannot make 'tmlib' your current working directory (see Chapter 1 and its worksheet for instructions), additional path specifiers may be required in the read command. With the Pauli-package procedures, Pauli cliffors can be represented as the sum of scalars and three-dimensional vectors (arrays). The Cartesian unit vectors are predefined to be e1, e2, e3; the unit element is e0 = 1. The algebraic product is indicated by the binary operator &v, and the command evalPa evaluates a linear combination of Pauli elements. Try, for example,

> vec1 := array([3,0,0]); vec2 := array([1,2,0]);

$$vec1 := [3\ 0\ 0]$$
$$vec2 := [1\ 2\ 0]$$

> evalPa(vec1 + 3*vec2);

$$6e1 + 6e2 \qquad (9.82)$$

> vec12 := vec1 &v vec2;

$$3 + [0\ 0\ 6I] \qquad (9.83)$$

The second command evaluates the Pauli cliffor indicated, namely the sum vec1 + 3 vec2, and the third command computes the algebraic product of the two real vectors. The scalar part of the result is also real and gives the dot product of the two vectors, whereas the vector part is imaginary and gives the area of the parallelogram whose sides are vec1 and vec2. Try also

> e1 &v e1;

$$1 + [0\ 0\ 0]$$

> e1 &v e2; e2 &v e1;

$$[0\ 0\ I]$$
$$[0\ 0\ -I] \qquad (9.84)$$

```
> e1 &v e2 &v e3;
```
$$I + [0\ 0\ 0]$$

```
> p := 1 + array([2,3,4]);
```
$$p := 1 + [2\ 3\ 4]$$

```
> evalPa(2 + 4*p + 3*e1);
```
$$6e0 + 11e1 + 12e2 + 16e3] \qquad (9.85)$$

```
> r := t + x * e1 + y * e2 + z * e3):
> e1 &v r &v e1;
```
$$t + [x - y - z] \qquad (9.86)$$

The dot product of two cliffors p and q is given by p &dot q. In particular, p &dot e.n gives the n-th component of p:

```
>  for mu from 0 to 3 do r &dot e.mu od;
```

$$
\begin{array}{c}
t \\
x \\
y \\
z
\end{array} \qquad (9.87)
$$

The spatial reversal, hermitean conjugate, inverse, and square norm of an element p are given respectively by Pbar(p), Pdag(p), Pinv(p), Pnorm(p). Thus

```
> Pbar(r);
```
$$t - [x,\ y,\ z\ ]$$

```
> Pbar(r) &v r;
```
$$t^2 - x^2 - y^2 - z^2 + [0,0,0]$$

```
> Pnorm(r);
```
$$t^2 - x^2 - y^2 - z^2 \qquad (9.88)$$

The command Pmat(p) represents the element p in standard 2×2 matrix form, whereas matP reverses the process. The Pauli expression for the rotation by the vector angle theta is given by rotate(theta), and the boost by the vector boost parameter w is given by boost(w):

> assume(theta >= 0): rotate(theta &v e2);

$$\cos(\frac{1}{2}\theta\tilde{\,}) + [0 \ -I\sin(\frac{1}{2}\theta\tilde{\,})\,0] \tag{9.89}$$

> Lambda := boost(3*e1) &v rotate(Pi*e2/3);

$$\Lambda := \frac{1}{2}\cosh\left(\frac{3}{2}\right)\sqrt{3} + \left[\frac{1}{2}\sqrt{3}\sinh\left(\frac{3}{2}\right) \ -\frac{1}{2}I\cosh\left(\frac{3}{2}\right)\frac{1}{2}\sinh\left(\frac{2}{3}\right)\right] \tag{9.90}$$

The symbol $\theta\tilde{\,}$ means θ with assumed properties.

The rotation and boost are special examples of functions of Pauli cliffors. The more general expression for the function f of an element p is given by Pfun(f,p). The commands combine, convert, simplify, normal, eval, evalc, evalf, expand, and factor can be implemented on cliffors in the same way, and optional parameters can be added in additional arguments:

> Pfun(evalf,Lambda);

$$2.037246487 + [1.844010100 \ -1.176204808I \ 1.064639728] \tag{9.91}$$

> Pfun(sin,Pi*e3/3);

$$\left[0 \ 0 \ \frac{1}{2}\sqrt{3}\right] \tag{9.92}$$

Finally, an arbitrary non-null Pauli cliffor can be expressed as the product of a normalization factor, a rotation, and a boost: use the procedure Pfactor for this purpose. It returns a list of the normalization factor, the vector rotation angle, and the vector boost parameter. If the cliffor is null, it is proportional to a projector whose direction is given by the boost. Pfactor in this case returns a zero normalization factor, the boost parameter, and the rotation angle. A complete list of the available commands is given in Section 9.10. The procedures in the directory Pauli are in ASCII source code to ensure maximum independence of the computer platform and release number. They can be viewed and modified in any ASCII editor, and if you want, they can be saved in internal Maple format for speedier execution. More applications are given in the worksheet for this chapter.

9.9 Problems

1. Let $\mathbf{r}$ be the vector for a point in space relative to a fixed origin, β be a real constant scalar, and $\mathbf{a}, \mathbf{b}$ real orthogonal constant vectors: $\mathbf{a} \cdot \mathbf{b} = \mathbf{0}$.

   (a) Show that the equation $\mathbf{r} \cdot \mathbf{b} = \beta$ determines a plane perpendicular to $\mathbf{b}$. That is, show that any vector joining two points in the plane is perpendicular to $\mathbf{b}$.

   (b) Find the location of the point in the plane which is closest to the origin.

   (c) Demonstrate that the equation $\mathbf{r} \times \mathbf{b} = \mathbf{a}$ determines a line parallel to $\mathbf{b}$. That is, show that any vector joining two such points is aligned with $\mathbf{b}$.

   (d) What point in the line in part (c) is closest to the origin?

   (e) Show that the position $\mathbf{r}$ satisfying $\mathbf{rb} = \beta + i\mathbf{a}$ is the point of intersection of the plane of part (a) and the line of part (c). Find this position. [Hint: multiply the algebraic condition for $\mathbf{rb}$ by $\mathbf{b}^{-1}$.]

2. Let $p = p_0 + \mathbf{p}$ be a Pauli cliffor, the square of whose vector part does not vanish: $\mathbf{p}^2 \neq 0$. Define the elements $P_+ := \frac{1}{2}(1 + \hat{\mathbf{p}})$ and $P_- = \frac{1}{2}(1 - \hat{\mathbf{p}}) = \bar{P}_+$ where $\hat{\mathbf{p}} = \mathbf{p}/\sqrt{\mathbf{p}^2}$ is the (possibly complex) unit vector parallel to $\mathbf{p}$.

   (a) Show that the cliffors are *idempotent*: $P_+^2 = P_+$, $P_-^2 = P_-$.

   (b) Show also that they are *orthogonal* $(P_+P_- = 0)$ and *complementary* $(P_+ + P_- = 1)$.

(c) Prove that the idempots $P_\pm$ obey eigenvalue equations of the form

$$pP_\pm = \epsilon_\pm P_\pm \qquad (9.93)$$

where the eigenvalues $\epsilon_\pm$ are scalars, and find $\epsilon_\pm$.

(d) Show that any power p^n of p obeys

$$p^n = [p\,(P_+ + P_-)]^n = \epsilon_+^n P_+ + \epsilon_-^n P_-. \qquad (9.94)$$

(e) Prove that any analytic function $f\,(x)$ can be written as a function of p given by

$$f\,(p) = \frac{1}{2}\,[f\,(\epsilon_+) + f\,(\epsilon_-)] + \frac{\hat{P}}{2}\,[f\,(\epsilon_+) - f\,(\epsilon_-)]\ , \qquad (9.95)$$

and that in the limit $\mathbf{p}^2 \to 0$, the expression becomes

$$f\,(p) = f\,(p_0) + f'\,(p_0)\,\mathbf{p},\ \mathbf{p}^2 = 0 \qquad (9.96)$$

where $f'\,(x) = df\,(x)\,/dx$.

3. Apply the results of Problem 2 to the inverse and square root functions of an invertible Pauli element p. Show that they give

$$p^{-1} = \frac{\bar{p}}{p\bar{p}} \qquad (9.97)$$

and

$$p^{1/2} = \frac{p + \sqrt{p\bar{p}}}{\sqrt{2}\,(p_0 + \sqrt{p\bar{p}})}. \qquad (9.98)$$

You may find the following relations useful (you will want to verify them!):

$$\sqrt{a+b} \pm \sqrt{a-b} = \left[2\left(a \pm \sqrt{a^2 - b^2}\right)\right]^{1/2} = \left[\frac{2b^2}{a \mp \sqrt{a^2 - b^2}}\right]^{1/2}. \qquad (9.99)$$

4. Find the rotation operator of the form $\exp\,(-i\boldsymbol{\theta}/2)$ which is equivalent to a rotation first about the e_2-axis by $\pi/2$ and then about the e_3-axis by the same angle. Compare it to the rotation operator when the 90-degree rotations are performed in the opposite order.

5. Express the acceleration $\ddot{\mathbf{r}}'$ in terms of the position vector $\mathbf{r}$ and its time derivatives in a frame rotating at constant angular velocity $\boldsymbol{\omega}$. (See Section 9.4.5.) Show that an object of mass m in the rotating frame experiences two fictional forces, namely the *centrifugal force* $m\boldsymbol{\omega}\times(\mathbf{r}\times\boldsymbol{\omega})$ and the Coriolis force $2m\dot{\mathbf{r}}\times\boldsymbol{\omega}$. Verify that the centrifugal force is directed radially outward from the rotation axis, whereas the Coriolis force encourages a clockwise circulation of streams in the northern hemisphere (by definition, the angular velocity of the rotating frame points towards the "north").

9.10 Chapter Summary

9.10.1 New Concepts

scalars, vectors, pseudovectors, pseudoscalars

axial vectors

cliffors

real, imaginary, scalar, and vector parts of element

dot and cross products, the algebraic product

bivectors, trivectors

field, algebra, subalgebra, division algebras

the center of an algebra

Pauli spin matrices

quaternions

involutory transformations, automorphisms, anti-automorphisms

spatial inversion, spatial reversal, hermitean conjugation

4-vectors and 6-vectors, paravectors and parabivectors

Minkowski spacetime, the Minkowski metric

Levi-Civita symbol

geometric (Clifford) algebras

the Pauli algebra

Hodge duals

fictional forces: centrifugal and Coriolis forces

rotations, boosts, Lorentz transformations

idempots, eigenelements

9.10.2   Pauli Package commands

&v, &dot (algebraic and dot products)

e0, e1, e2, e3 (basis vectors)

Pfactor(p) (factor cliffor p into *norm* ∗ *rotation* ∗ *boost*)

boost(w), rotate(theta) (boost and rotation)

evalPa(p) (evaluate Pauli cliffor p as scalar + vector)

evalPe(p) (evaluate cliffor p as list [scalar, vector])

Pscal(p) (find scalar part of cliffor p)

Pvec(p) (find vector part of cliffor p)

Preal(p)(find real part of cliffor p)

Pimag(p)(find imaginary part of cliffor p)

Pnorm(p) (find square norm of cliffor p)

Pbar(p) (find spatial reverse of cliffor p)

Pdag(p) (find hermitean conjugate of cliffor p)

Pinv(p) (find inverse of cliffor p)

Pmat(p) (find standard 2×2 matrix representation of p)

matP(m) (convert 2×2 matrix to cliffor as scalar + vector)

Pfun(f,p[,o]) (find $f(p)$ where f is any Maple function and [o] are options)

Pdi(f,x) (calculate the Pauli gradient $\partial f(x)$)

Pdib(f,x) (calculate the Pauli gradient $\bar{\partial} f(x)$)

Appendix A

Greek Alphabet

The physical sciences make extensive use of the Greek alphabet. Since at least parts of this alphabet may be new to some readers, and because the way the lower-case letters are drawn can vary, we list in the table A.1 the upper- and lower-case characters and their names. No attempt is made to list pronunciations since it is common to mix Greek and English.

Lower	Upper Case	Name	Lower	Upper Case	Name
α	A	alpha	ν	N	nu
β	B	beta	ξ	Ξ	xi
γ	Γ	gamma	o	O	omicron
δ	Δ	delta	π	Π	pi
ϵ, ε	E	epsilon	ρ	P	rho
ζ	Z	zeta	σ	Σ	sigma
η	H	eta	τ	T	tau
θ, ϑ	Θ	theta	υ	Y	upsilon
ι	I	iota	ϕ, φ	Φ	phi
κ	K	kappa	χ	X	chi
λ	Λ	lambda	ψ	Ψ	psi
μ	M	mu	ω	Ω	omega

Table A.1. The Greek Alphabet

Appendix B

Units

The following is a printout of the ASCII file 'units'. Use it to determine the abbreviations of the constants and symbols. The user can add more constants and conversion factors with any ASCII editor.

```
# LIST OF UNITS AND CONSTANTS
#
# converts expressions automatically to MKSA (SI) units
# prepared by W. E. Baylis for use in his text "Theoretical Methods in the
#     Physical Sciences, An Introduction to Problem Solving with MAPLE V",
#     Birkhauser Boston 1994.
# based on E. R. Cohen and B. N. Taylor, Rev Mod Phys 59, 1121 (1987)
#
#  Metric prefixes
Exa     := 10^18    :
Peta    := 10^15    :
Tera    := 10^12    :
Giga    := 10^9     :
Mega    := 10^6     :
kilo    := 10^3     :
milli   := 10^(-3)  :
micro   := 10^(-6)  :
nano    := 10^(-9)  :
pico    := 10^(-12) :
femto   := 10^(-15) :
atto    := 10^(-18) :
#
```

```
#Common conversions
deg_    := Pi*rad_/180      :#angular measure
cycle_  := 2*Pi*rad_        :# relation between cycle and radian
rad_    := 1                :# radian (caution: some authors use cycle_=1)
Hz_     := cycle_/s_        :# Hertz (frequency)
hr_     := 60*min_          :
min_    := 60*s_            :
day_    := 24*hr_           :
yr_     := 365.24219*day_   :#tropical year
mm_     := milli*m_         :
cm_     := 0.01*m_          :
km_     := kilo*m_          :
in_     := 2.54*cm_         :
ft_     := 12*in_           :
yd_     := 3*ft_            :
mi_     := 5280*ft_         :#statute mile
ly_     := 94605282*Giga*m_ :#light-year
pc_     := 3.261633*ly_     :#parsec
hectare_ := (100*m_)^2      :# 100 ares, unit of area
acre_   := 0.40468564*hectare_:#acre, 160 sq.rods
l_      := m_^3/1000        :# liter
L_      := l_               :# alternative liter symbol
ml_     := milli*l_         :
cc_     := ml_              :# cubic centimeter
gal_    := 3.785412*l_      :# gallon (US, liquid)
gal_imp := 4.54609*l_       :# Imperial gallon
qt_     := gal_/4           :# quart(US, liquid)
pt_     := qt_/2            :# pint(US, liquid)
g_      := kg_/1000         :# gram (cgs unit of mass)
Tonne_  := 1000*kg_         :# metric ton
lb_     := 0.453592*kg_     :# pound mass (do not confuse with lb_force)
ton_    := 2000*lb_         :# short ton
oz_     := lb_/16           :# ounce (US, fluid)
N_      := kg_*m_/s_^2      :# Newton (metric unit of force)
dyne_   := g_*cm_/s_^2      :# cgs unit of force
lb_force := 4.44822*N_      :# caution: lb is used both for force and mass
P_      := g_/(cm_*s_)      :# poise, unit of viscosity
J_      := N_*m_            :# Joule (metric unit of energy)
erg_    := dyne_*cm_        :# cgs unit of energy
cal_    := 4.1868*J_        :# calorie (cgs heat unit)
Cal_    := kilo*cal_        :# dieticians calorie
Btu_    := 1055.056*J_      :# British thermal unit
eV_     := 1.60217733*10^(-19)*J_:# electon Volt (energy)
Hartree_:=27.21139613*eV_  :# atomic unit of energy = 2 Rydbergs of energy
W_      := J_/s_            :# Watt
hp_     := 745.7*W_         :# horsepower
```

```
Pa_     := N_/m_^2          :# Pascal, the metric unit of pressure
atm_    := 101325*Pa_       :# atmosphere = 29.9213 in_ of mercury
bar_    := 100000*Pa_       :# bar of pressure
torr_   := atm_/760         :# mm_ height of mercury
V_      := J_/C_            :# metric unit of Volt
StatV_  := 299.792458*V_    :# cgs (esu) unit of potential
C_      := A_*s_            :# Coulomb (metric unit of electric charge)
StatC_  := C_/2997924580    :# cgs (esu) unit of electric charge
StatA_  := A_/2997924580    :# cgs (esu) unit of current
F_      := C_/V_            :# Farad (metric unit of capacitance)
Wb_     := V_*s_            :# Weber (metric unit of magnetic flux)
H_      := V_*s_/A_         :# Henry (metric unit of inductance)
ohm_    := V_/A_            :# metric unit of resistance
T_      := Wb_/m_^2         :# Tesla (metric unit of magnetic field B)
gauss_  := T_/10000         :# cgs (emu) unit of magnetic induction B
Oe_     := gauss_           :# oersted, cgs (emu) unit of magnetic field H
#
# Physical Constants
mol_    := 6.0221367*10^(-23)    :#Avogadro constant, 1 mole
alpha_  := 1/137.036             :#fine-structure constant
c_      := 299792458*m_/s_       :#speed of light (exact)
h_      := 6.6260755*10^(-34)*J_*s_:#Planck constant
hbar_   := 1.05457266*10^(-34)*J_*s_:#Planck constant/2 Pi
e_      := 1.60217733*10^(-19)*C_:#elementary charge
Ryd_inf:= 10973731.57/m_         :#Rydberg constant for infinite mass
mu_B    := 9.2740154*10^(-24)*J_/T_:#Bohr magneton
mu_0    := 4*Pi*10^(-7)*N_/A_^2  :#permeability of vacuum
eps_0   := 1/(muO_*c_^2)         :#permittivity of vacuum
G_      := 6.6726*10^(-11)*m_^3/(kg_*s_^2):#Newtonian gravitational constant
k_      := 8.31451*J_/(K_*mol_)  :#Boltzmann constant
m_e     := 9.1093897*10^(-31)*kg_:#electron mass
m_p     := 1836.1527*m_e         :#proton mass
m_n     := 1838.68366*m_e        :#neutron mass
u_      := 1.66054*10^(-27)*kg_  :#unified atomic mass unit (amu) M(C12)/12
m_sun   := 1.989*10^30*kg_       :#mass of sun
m_earth:= 5.98*10^24*kg_         :#mass of earth
m_moon  := 7.34*10^22*kg_        :#mass of moon
bohr_   := 0.52917725*10^(-10)*m_:#bohr radius a0 (atomic unit of length)
R_moon  := 1.74*Mega*m_          :#average radius of moon
R_earth:= 6.37*Mega*m_           :#average radius of earth
R_sun   := 696*Mega*m_           :#average radius of sun
r_earth:= 149.59787*Giga*m_      :#average radius of earth's orbit about sun
AU_     := r_earth               :#Astronomical Unit
r_moon  := 384.0*Mega*m_         :#average radius of moon's orbit about earth
g_acc   := 9.80665*m_/s_^2       :#acceleration of gravity on earth's surface
```

Appendix C

Constants and Functions in Maple

We summarize below the names of a few common constants which Maple knows.

Constant	Value
Pi	$\pi = 3.14159265...$
E	$e = \exp(1) = 2.718281728459...$
I	$i = \sqrt{-1}$
gamma	$\gamma = 0.5772156649...$
infinity	∞

The following table lists some of the common functions available in Maple. While some of these functions are in the Maple core, others are available in the Maple library. All those listed here can be invoked directly, without calling any special packages.

Function	Description
abs(x)	$\|x\|$
sqrt(x)	$\sqrt{x}$
sin(x)	sine of x
cos(x)	cosine of x
tan(x)	tangent of $x = \sin(x)/\cos(x)$
cot(x)	cotangent of $x = 1/\tan(x)$
csc(x)	cosecant of $x = 1/\sin(x)$
sec(x)	secant of $x = 1/\cos(x)$
arccsin(x)	inverse function of sine of x
arctan(x)	inverse function of tangent of x
exp(x)	exponential e^x
sinh(x)	hyperbolic sine of x, $(e^x - e^{-x})/2$
cosh(x)	hyperbolic cosine of x, $(e^x + e^{-x})/2$
tanh(x)	hyperbolic tangent of x, $\sinh(x)/\cosh(x)$
coth((x)	hyperbolic cotangent of x, $1/\tanh(x)$
ln(x)	natural logarithm of x
log10(x)	common logarithm (base 10) of x
n!	n factorial, $n(n-1)(n-2)\cdots(2)(1)$
GAMMA(x)	Gamma function, $\Gamma(x) = \int_0^\infty dt\, t^{x-1} \exp(-t)$ $\Gamma(n+1) = n!$ for integer n
Beta(x,y)	Beta function $B(x,y) = \int_0^1 dt\, t^{x-1}(1-t)^{y-1}$ $= \Gamma(x)\Gamma(y)/\Gamma(x+y)$
binomial(n,m)	binomial coefficient, $\binom{n}{m} = \frac{n!}{m!(n-m)!}$
rand	random number
erf(x)	error function, $\frac{2}{\sqrt{\pi}}\int_0^x ds\, e^{-s^2} = \frac{1}{\sqrt{2\pi}}\int_{-\sqrt{2}x}^{\sqrt{2}x} dt\, e^{-t^2/2}$
erfc(x)	complementary error function, $1 - \mathrm{erf}(x)$
Dirac(x)	Dirac delta function, $\delta(x)$

Index